FOOD BIOTECHNOLOGY: CURRENT ISSUES AND PERSPECTIVES

FOOD BIOTECHNOLOGY: CURRENT ISSUES AND PERSPECTIVES

SARAH ELDERIDGE
EDITOR

Nova Science Publishers, Inc.
New York

Senior Editors: Susan Boriotti and Donna Dennis
Coordinating Editor: Tatiana Shohov
Office Manager: Annette Hellinger
Graphics: Wanda Serrano and Matt Dallow
Editorial Production: Maya Columbus, Alexis Klestov, Vladimir Klestov,
Matthew Kozlowski and Lorna Loperfido
Circulation: Ave Maria Gonzalez, Vera Popovic, Sean Corkery, Raymond Davis,
Melissa Diaz, Magdalena Nuñez, Marlene Nuñez and Jeannie Pappas
Communications and Acquisitions: Serge P. Shohov
Marketing: Cathy DeGregory

Library of Congress Cataloging-in-Publication Data
Available Upon Request

ISBN: 1-59033-848-0.

400 Oser Ave, Suite 1600
Hauppauge, New York 11788-3619
Tele. 631-231-7269 Fax 631-231-8175
e-mail: Novascience@earthlink.net
Web Site: http://www.novapublishers.com

Printed in the United States of America

CONTENTS

PREFACE

Chapter 1 provides basic information on the science of biotechnology. It discusses regulatory policies and issues of concern about the use of biotechnology to modify foods through genetic engineering. It describes the scientific processes used and current products available. It explains how all three major federal agencies- the Food and Drug Administration, the U.S. Department of Agriculture, and the Environmental Protection Agency-regulate these foods. Consumers have expressed concerns about the uncertain long-term impact on public health and the environment, particularly the consequences of cross pollination, the impact on ecosystems, and the development of resistance with the use of some bioengineered plant pesticides. Some critics also question the expanding market power of a few multinational companies, the growing unease in international trade relationships over the fast adoption by U.S. farmers of bioengineered crops, and the current federal structure by which it is regulated. Others argue that food biotechnology will enhance crop yields, produce foods with novel characteristics, while using fewer pesticides.

The European Union (EU) is the second largest market for U.S. agricultural exports. The EU's ban on meat produced using growth-promoting hormones is a food safety issue that has been particularly contentious in U.S.-E.U. agricultural trade relations. Chapter 2 discusses this as well asEU policy on bio-engineered products.

StarLink(tm) is a corn variety that has been genetically modified to contain an insecticidal protein derived from a naturally occurring bacterium (bacillus thuringiensis, or Bt). Several bills were introduced in the 106th Congress to require labeling of foods containing genetically modified organisms (GMOs). In addition, a bill was introduced at the end of the 106th Congress (S. 3184) to amend the Federal Food, Drug and Cosmetic Act to require pre-market consultation and approval for approval for foods containing GMOs. These issues are covered in chapter 3.

Congressional interest in the labeling of genetically modified foods (GM foods) has been rising. Congress took no action on the two bills introduced into the 106th Congress to mandate labeling of all genetically modified crops (GM crops) and foods, but will likely debate other proposals in the 107th Congress. The Food and Drug Administration (FDA) issued its current GM food labeling policy in May 1992. At that time, the agency determined that it would regulate GM foods no differently than foods created by conventional means because FDA considered them substantially equivalent to traditional foods and decided that no special label would be needed. In the intervening period, extensive public debate surrounding the genetic modification of foods has led some consumers to call for labeling of such products. On January 18, 2001, FDA published a draft guidance for industry on voluntary efforts to label GM foods. A label would permit consumers to choose to avoid purchasing or consuming them. At the same time, the agency proposed a rule that would make mandatory the current voluntary consultation process whereby the industry would have to formally notify the agency 120 days before releasing a food product onto the market and would have to supply certain safety test data to FDA. The federal government's role in regulating these foods is explained in chapter 1. Many oppose labeling because to make such labels "truthful and not misleading" all commodities would need to be segregated and tested, and the label would not have room to impart information that could not be distributed in other ways. Chapter 4 focuses specifically on views surrounding the labeling of GM foods.

Since the first genetically engineered (GE) crops became commercially available in the mid-1990s, U.S. farmers have been rapidly adopting them in efforts to lower production costs and raise crop yields. A so-called second generation of GE commodities could shift the focus of agricultural biotechnology from the input side (i.e. farm production benefits) to the output side (i.e. consumer benefits). Even though only 12 different plants with GE traits have been approved for commercial use, at least 60 percent of all U.S. foods likely contain at least some GE material. That's largely because two of those plants—corn and soybeans, where farmers have widely adopted GE varieties—are used in many different processed foods. Chapter 5 covers this issue.

Chapter 6 provides background on "terminator genes" in seeds and briefly examines three issues associated with these new types of biotechnologies: whether they influence vertical integration in the seed industry, whether research and development of terminator-type genes should be funded by the U.S. government under Cooperative Research and Development Agreements (CRADAs), and whether food security and biodiversity of developing countries could be impacted.

The United States and several other countries in May 2003 initiated a formal challenge before the World Trade Organization (WTO) of the European Union's (EU's) de facto moratorium on approving new agricultural biotechnology products, in effect since 1998. The moratorium reportedly has cost U.S. corn growers some $300 million in exports to the EU annually. Chapter 7 explores this issue.

The Biosafety Protocol to the 1992 Convention on Biological Diversity, adopted in early 2000, addresses the safe handling, transfer, and trade of biological organisms. The Protocol sets forth procedures and rules concerning trade in biological products, including genetically modified organisms (GMOs) that have engendered controversy, especially when they are used as agricultural crops. These rules are of key importance to U.S. economic interests in agriculture as well as those dealing in other genetically modified organisms. Chapter 8 provides a summary of the key provisions of the Protocol and the major issues associated with them.

Chapter 9 discusses this issue that currently, no legislative proposals requiring mandatory labeling of genetically engineered foods have been introduced in the 107th Congress, as was the case in the 106th. An important thrust behind current Congressional proposals is the support of research and development programs in agricultural biotechnology.

In Chapter 10 genetically engineered (GE) wheat varieties developed by the biotech industry and university researchers are discussed. They hold considerable promise for producers. Market acceptance focuses on its direct acceptance by consumers. Most research suggests GE foods are safe to eat; however, uncertainty about health and environmental effects has led to public opposition particularly in Europe and Japan. While developers of the technologies are seeking regulatory approval, they indicate that they will postpone their commercialization until market acceptance is assured and systems for separating GE and non-GE wheat are in place.

Commercial use of roundup ready soybeans in Brazil remains illegal despite apparent widespread planting. A 1998 government approval of their commercial use remains suspended by court injunction, and resolution over their commercial legality is being considered by an appellate court. Chapter 11 discusses the outcome on commercial legalization of genetically engineered crops in Brazil. It may have important consequences for intellectual property rights, as well as for international trade in genetically engineered crops.

Chapter 12 talks about the differences between the United States and its trading partners over genetically engineered (GE) crops and food products that contain them pose a potential threat to, and in some instances have already disrupted, U.S. agricultural trade. Underlying the conflicts are pronounced differences between the United States and several important trading partners in consumer attitudes about GE products and their potential health and environmental effects.

U.S. farmers have been rapidly adopting genetically engineered (GE) crops—mainly corn, soybean, and cotton varieties—to lower production costs and improve management. However, the U.S. agricultural economy is highly dependent upon exports, at a time when many foreign consumers are wary of the products of agricultural biotechnology. As a result, U.S. exporters often have encountered barriers to trade in these markets. Chapter 13 discusses these issues.

Chapter 1

FOOD BIOTECHNOLOGY IN THE UNITED STATES: SCIENCE, REGULATION, AND ISSUES*

Donna U. Vogt and Mickey Parish

INTRODUCTION

Genetic modification of agricultural crops promises the availability of food products with more desirable traits, such as higher quantities of vitamins or lowered amounts of saturated fats for consumers, reduced use of pesticides and other chemicals for environmentalists, and increased yields for growers. Traditional plant breeding, the conventional method to modify plants' genes, has produced similar benefits. But recent biotechnological innovations allow scientists to select specific genes from one plant or animal and introduce them into another to confer desirable traits. This produces the new plant or animal more quickly than conventional methods, and creates plants and animals with traits not found previously in nature. Proponents argue that advances in genetics and new technologies can produce foods with greater yields to feed the growing world population in the 21st century. Critics are concerned that this technology produces uncertainties about potential long-term impacts on public health and the environment, and increases problems related to trade.

---

* Excerpted from CRS Report RL30198.

The federal government under statute and through regulation attempts to ensure that food manufacturers produce safe products. However, some Members of Congress have asked whether federal regulations adequately manage genetic engineering risks to public health and safety, and the environment. Nine bills introduced into the 106th Congress asked that the Administration pay more attention to these risks. Similar bills are likely to be introduced into the 107th Congress which will call for the same additional scrutiny and research.

This chapter discusses the science of food biotechnology, and the federal structure by which it is regulated. Because U.S. farmers are adopting this technology at a rapid rate, some observers advocate a more active role for the federal government to ensure that farmers have equal access to this technology. Others believe that federal officials should play a more active role in protecting the environment, funding more research, and participating in international trade negotiations to ensure that trade continues to expand for genetically engineered crops. Trading partners often label food products that have been genetically modified as genetically modified organisms (GMOs). Many of those partners have labeling requirements for GMOs to allow consumers the "right to know" their food content.

Several congressional committees oversee federal governance of genetically engineered foods and biotechnology. In the Senate, food biotechnology issues are considered by the Committees on Agriculture, Nutrition, and Forestry; Health, Education, Labor and Pensions; Environment and Public Works; and Governmental Affairs. In the House, food biotechnology issues are considered by the Committees on Agriculture; Energy and Commerce; Government Reform; and Science. The Appropriations Committees of both the House and Senate have oversight responsibility on how the major federal agencies set and enforce policies affecting the safety of genetically engineered foods.

THE SCIENCE OF GENETIC ENGINEERING

Biotechnology is defined as the use of biological processes for the development of products such as foods, enzymes, drugs, and vaccines. Biotechnology is the new label for a process that humans have used for thousands of years to ferment foods such as beer, wine, bread, and cheese. In these cases, biological processes are used to alter raw food products to produce more stable foods. Presently, the term biotechnology is used to describe genetically engineered foods that contain genes modified by modern technologies.

When plants breed in the wild, genetic changes occur spontaneously and result in a haphazard transfer of a large number of genes within closely related species. Traditional plant breeding is more selective and creates plants with improved yields or some other desirable trait among closely related species. Often, unwanted genes conferring undesired characteristics may be transferred along with the desired characteristics. Breeding itself takes time due to the need to backcross each plant to eliminate undesirable traits.

Modern genetic engineering gives greater control of the process and transfers specific genetic material into the cells of a plant. This method can reduce the likelihood of unexpected results. Also, plant breeders can use the newer genetic techniques to move genes among unrelated species to yield plants with novel traits that could not be produced by traditional breeding.

Several techniques are employed by genetic engineers. All involve DNA transfers from one plant or animal to another. DNA, deoxyribonucleic acid, is the chemical from which genes are constructed. Specialized laboratory techniques, generally referred to as recombinant DNA (rDNA) techniques are used to manipulate DNA isolated from animal, plant, or microbial cells and to introduce the engineered DNA sequences into another organism. The laboratory techniques of genetic engineering may involve direct cellular uptake of DNA, forced introduction of DNA into a cell, or the use of a non-pathogenic carrier to transmit genetic material into a cell. Additionally, some microbial cells in immediate contact can transfer DNA directly from one cell to another. Plants may also be genetically modified by fusion of whole cells.[1]

After plant cells are genetically modified, tissue culture techniques are used to encourage growth of the modified cells into whole plant systems with leaves, stems, and roots. New food plant characteristics depend on which genes are transferred, whether these genes are switched on (expressed), and the interaction between genes and the cellular environment in which they reside.

Use of Biotechnology to Produce Food

The first wave of agricultural biotechnology food products is not substantially different from those foods already familiar and available to consumers. These modified agricultural commodities have, for the most part, directly benefitted agricultural producers with increased yields and reduced production costs.

---

[1] Grosser, J., and F. Gmitter. "Protoplast Fusion And Citrus Improvement." *Plant Breeding Reviews*, (Ed., J. Janick), v. 8, Chap.10. Timer Press, Portland, OR.,1990.

According to an industry trade association, genetically engineered food crops planted and marketed by U.S. farmers include corn, canola, rice, tomatoes, potatoes, and soybeans. Peppers, sunflowers and peanuts are in the pipeline for approval. Other genetically engineered food crops, such as sugar beets, wheat, squash, papayas, berries, bananas, and pineapples, have been developed in laboratories, and will go through the approval process for marketing within the next few years. Non-food plants that are being genetically modified include trees, for pulp wood, and cotton, although cotton seed oil may be used in food products.

Genetically Modified Whole Food Products

Some of the products referred to above have been genetically modified to be either resistant to pesticides,[2] or to make their own pesticides. Those modified to resist pesticides allow farmers to use a herbicide for weed control without killing the food crop. This use reduces competition from weeds for nutrients and increases yields. Other products of bioengineering produce insecticides within their cells. Crops such as corn, cotton, and potatoes, have been genetically engineered to make their own pesticide.

Farmers rapidly accepted these genetically engineered field crops with increasing acreage of herbicide resistant crops. The use of these plants reduces the need to plow, decreases the amount of chemical herbicide needed, produces higher yields, and can deliver a cleaner and higher grade of grain and product.[3] Of the total 1998 crop, approximately 25% of planted corn was genetically modified, 38% of planted soybeans, 45% of cotton, and 42% of canola.[4] In 1999 the percentage of acres planted with corn grew to 37% but declined to 25% in 2000. In 1998 in the United States, herbicide tolerant soybeans became the dominant bioengineered crop, with 4.1 million acres planted (about 36% of the total U.S.

---

[2] The term "pesticide" is the general category of substances that are toxic to pests; herbicides, insecticides, and rodenticides are all pesticides.

[3] James, Clive. "Global Review of Commercialized Transgenic Crops." *International Service for the Acquisition of Agri-biotech Applications Briefs*, no. 8, 1998, ISAAA: Ithaca, NY.

[4] Statistics come from the following sources: Terri Dunahay. Economic Research Service, January 20, 1999; A.S. Moffat. "Toting Up the Early Harvest of Transgenic Plants." *Science*, v. 282. p. 2176-2178; M. Pollan. "Playing God in the Garden." *New York Times Magazine*, October 25, 1998. p. 44-51, 62-63, 92-93; R. Ortiz. "Critical Role of Plant Biotechnology for the Genetic Improvement of Food Crops: Perspectives for the New Millennium." *Electronic Journal of Biotechnology*, v. 1, no. 3, December 15, 1998. [http://www.ejb.com]; J. Bernice. "Medicine Man." *Farm Journal*, mid-January 1998. p. 22; *Genetic Engineering News*, February 1, 1999: "Cancer Vaccine Made in a Plant-based System", p. 1 were collected from: *Science* (1998), v. 282, p. 2176; Michael Pollan, *N.Y. Times Magazine*, October 1998; R. Ortiz, *Electronic J. Biotechnology*, v. 1, no. 3, December 15, 1998. [http://www.ejb.com]; S.D. Moore. "Agro-chemical Rivalry Heats Up: AHP-Monsanto Pact Raises Pressure." *The Wall Street Journal Europe*, June 5-6, 1998, p. 10.

acreage planted in soybeans). In 1999, bioengineered soybean plantings increased to 47% and continued to increase to 54% in 2000. Cotton has followed a similar trend. Herbicide resistant crops of soybeans, corn, and cotton accounted for most of the acreage planted in biotech crops.

Various companies have targeted different genetic modifications to produce tomatoes that remain firm for a longer time and are reputedly more flavorful than traditional tomatoes. Calgene's Flavr Savr tomato was one of the first genetically engineered consumer-ready foods to receive federal approval for production and marketing in the United States (see text box on page 9). Another tomato engineered by Zeneca, used for production of a reduced-price tomato paste for sale in the United Kingdom, was withdrawn due to consumer protests.

Biotechnology is also used to produce experimental "transgenic" animals, in which the genetic material has been deliberately modified and to produce "clones" in which animals are reproduced artificially but the DNA is not modified. In agriculture, transgenic animals may be altered to produce higher yields of specific products (meat, milk, etc.) or to bring about commodities with enhanced characteristics, such as less cholesterol or reduced fat content. Although cloning has been used to reproduce animals for scientific purposes since the 1950s, its usefulness for the reproduction of identical livestock animals was only recently investigated. In 1995, sheep were cloned from embryonic cells in Scotland. In 1996, a substantial breakthrough followed when a sheep, Dolly, was cloned from an adult, nonembryonic cell.[5] Japanese scientists are creating high-value beef cattle through cloning. They have successfully cloned at least 19 calves from adult bovine cells. Because the cost of some premium beef roasts can be between $100 and $200 per pound in Japan, the Japanese cattle industry can support the expense of cloning prize beef cattle.[6] But even with those prices, the cost of genetically engineering cattle on a large scale could be prohibitive.

At this time, most of the research on transgenic animals in agriculture is experimental. The Biotechnology Industry Organization (BIO), an industry trade organization, estimates that the only transgenic animals that will be marketed within the next 6 years are transgenic fish that can grow to market size more rapidly than traditional farm-raised fish. Experiments are also being conducted to produce transgenic animals that yield human pharmaceuticals such as vaccines, growth hormones, blood-clotting factors, monoclonal antibodies and other drugs.

---

[5] Wilmut, I., A. E. Schnieke, J. McWhir, A.J. Kind, and K.H.S. Campbell. "Viable Offspring Derived from Fetal and Adult Mammalian Cells." *Nature*, v. 385, February 1997. p. 810-813.

[6] Normile, Dennis. "Bid for Better Beef Gives Japan a Leg Up on Cattle." *Science*, December 11, 1998. p. 1975-1976.

Enhancement of milk from cows, goats and sheep to contain these drugs is the target of much of this research.[7]

Animal products also have been changed by biotechnology. A food processing agent, chymosin (also called rennin) is an enzyme required to manufacture cheese. It was the first genetically engineered food additive to be used commercially. Traditionally, processors obtained chymosin from rennet, a preparation derived from the fourth stomach of milk-fed calves. Scientists engineered a non-pathogenic strain (K-12) of *E. coli* bacteria for large-scale laboratory production of the enzyme. This microbiologically produced recombinant enzyme, identical structurally to the calf derived enzyme, costs less and is produced in abundant quantities. Today about 60% of U.S. hard cheese is made with genetically engineered chymosin.[8] In 1990, FDA granted chymosin "generally-recognized-as-safe" (GRAS) status based on data showing that the enzyme was safe.[9] The final enzyme product is purified by removing potentially harmful substances, including the gene for antibiotic resistance used to engineer the microorganism that produces chymosin.

Bovine somatotropin (BST), also known as bovine growth hormone (BGH), occurs naturally in cows. When recombinant bovine somatotropin (rBST), is injected supplementally into dairy cattle, milk production may increase 10% to 15% (see text box). Genetically engineered microorganisms produce a consistent and affordable supply of this hormone as opposed to isolating the compound from limited bovine sources. According to an industry trade association, it is possible that up to 30% of U.S. dairy cows are injected with recombinant BST to increase milk yield.[10]

Future Products

Experts indicate that the "second wave" of genetically modified food products will target consumer and animal health issues and improve the nutrition content of

---

[7] Charles, Craig. "Near-term Benefits of Cloning Likely To Be Medical." *Washington Post*, March 29, 1997.

[8] BIO: Website of the Biotechnology Industry Organization, Washington, D.C. [http://www.bio.org/library/foodrep8.html].

[9] FDA classifies food additives into four categories: GRAS additives (generally- recognizedas- safe) are exempt from regulation because their extensive use has produced no known harmful effects; direct additives are intentionally added to foods; prior-sanctioned substances are substances which were approved by FDA or USDA before the passage of the 1958 food additives amendment to the FFDCA (Section 409, 21 U.S.C. 348); and indirect additives, sometimes called food contact substances, are often trace substances that leach from packaging materials and migrate to food during processing or storage.

[10] BIO: Website of the Biotechnology Industry Organization, Washington, D.C. [http://www.bio.org/whatis/foodwelcome.html].

certain foods. For example, vitamin A shortages that are a significant health concern in developing countries could be addressed by increasing the vitamin A content in canola oil. Genetically engineered soybeans, peanuts, and sunflowers may contain reduced levels of saturated fats or have altered fatty acid compositions for enhanced health benefits and to improve vegetable oil stability without the need for chemical hydrogenation. (Hydrogenation is used to create fats that are solid at room temperature — margarine, solid Crisco, etc.) Fruits may also be genetically engineered to contain vaccines. These fruits could be delivered without special care to developing countries which often don't have refrigeration necessary for current vaccines.[11] Genetically engineered salmon, trout, and flounder achieve market size in half the time of non-genetically engineered fish. Commercialization of such fish may help curtail over-fishing of native fish, as well as reduce consumer prices. In addition, scientists are using genetic engineering to make high-value products, such as special oils and chemicals. Rapeseed plants have been genetically modified to produce 35% more of a fatty acid, laurate, for use in soaps, detergents and other household items. Other plants and animals are being engineered to produce pharmaceuticals, specialty chemicals, and biologic agents.

FEDERAL RESPONSIBILITIES FOR REGULATING GENETICALLY MODIFIED FOODS

During the 1970s, the development of new techniques for transferring genes raised concerns about potential hazards. At the Asilomar Conference in February 1975, scientists working with this technology tried to reach a consensus to self-regulate research involving rDNA technology until its safety could be assured. The National Institutes of Health (NIH) became involved in 1976 when it published research guidelines using rDNA techniques. Until 1984, the NIH Recombinant DNA Advisory Committee was the primary federal entity that reviewed and monitored DNA research. However, a legal challenge forced the Reagan Administration to consider and propose policies to guide activities of federal agencies responsible for reviewing biotechnology research and its products.[12] In 1984, the White House Office of Science and Technology Policy

[11] Tacket, Carol O. "Vaccines from Edible Plants," paper presented at the 1998 FDA Science Forum, *Biotechnology: Advances, Applications and Regulatory Challenges.* Washington, December 8, 1998.

[12] *Foundation on Economic Trends* v. *Heckler*, 587 F. Supp 753 (D.C. 1984) 756 F.2d 143 (D.C. 1985).

(OSTP) published the "Coordinated Framework for Regulation of Biotechnology," a framework proposing that genetically engineered products would continue to be regulated according to their characteristics and novel features and not by their method of production. It also proposed that new biotechnology products be regulated under the existing web of federal statutory authority and regulation.[13]

In 1986, OSTP finalized this framework. The framework identified lead agencies to coordinate activities when and if jurisdictions overlapped. For example, the Food and Drug Administration (FDA) is responsible for regulating food and feeds in the market that have been modified through genetic engineering. The U.S. Department of Agriculture (USDA), Animal and Plant Health Inspection Service (APHIS), regulates importation, interstate movement, and environmental release of transgenic plants that contain plant pest components. It licenses, through permits, the field testing of food crops prior to commercial release. But agencies' responsibilities overlap as some plants have been modified to contain plant-pesticides. The Environmental Protection Agency (EPA) registers certain pesticides produced in transgenic plants prior to their distribution and sale and establishes pesticide tolerances for residues in foods.[14] APHIS and EPA together established that APHIS would regulate the plant itself, particularly those plants engineered to resist herbicides, and EPA would regulate the pesticide used on the herbicide resistant plant. They also together established procedures to review and approve field tests of modified plants and microorganisms. FDA has post-market authority to remove a food from the market. **Table 1** shows an overview of federal agencies' responsibilities.

---

[13] U.S. Congress. OTA. New Developments in Biotechnology — Field Testing Engineered Organisms, Genetic and Ecological Issues, no. 3, OTA-BA-350, May 1988. Washington, GPO, 1988. p. 60.

[14] EPA defines "plant-pesticides" as a "pesticidal substance that is intended to be produced and used in a living plant, or in the product thereof, and the genetic material necessary for the production of such a pesticidal substance." The DNA is not itself a pesticide. The plant uses the DNA to produce a pesticidal substance which is toxic to pests.

Table 1. Overview of Agency Responsibilities

Agency	Products Regulated	Reviews for Safety
FDA	Food, feed, food additives, veterinary drugs	Safe to eat
USDA	Plant pests, plants, veterinary biologic	Safe to grow
EPA	Microbial/plant pesticides, new uses of existing pesticides, novel microorganisms	Safe for the environment. Safety of a new use of a companion herbicide

Source: This data was compiled from tables found at the APHIS website: [http://www.aphis.usda.gov/biotech/OECD/usregs.htm]

Food and Drug Administration

The Federal Food, Drug, and Cosmetic Act (FFDCA) gives FDA broad authority to regulate foods by prohibiting the entry into interstate commerce of adulterated or misbranded foods. It is the legal responsibility of food manufacturers to produce foods that are not adulterated, unsafe, filthy, or produced under unsanitary conditions. FDA has authority to inspect foods and food facilities, both domestic and imported, to ensure that they are manufactured and held under acceptable conditions and are properly labeled. FDA can seize products or request that they be removed from the market if they do not meet federal requirements.

The Act also requires that "food additives" not be marketed unless they have received approval from FDA. But substances added to foods that are considered generally-recognized-as-safe or GRAS substances do not need agency approval. First articulated in the OSTP framework document in 1986, FDA determined in 1992 that bioengineered foods pose the kinds of scientific and regulatory issues that are not substantively different from those raised by non-bioengineered foods. Thus, FDA regulates foods that have been genetically modified or engineered no differently than foods created by conventional means.

In a May 1992 policy statement FDA described how its regulatory authority applies to new plant varieties and derived food products, including those developed through genetic engineering. The agency decided that companies developing genetically engineered foods would have to go through a special review in FDA only if:

- the gene transfer produces unexpected genetic effects;

- the levels of toxicants in the food are significantly higher than present in other edible varieties of the same species;
- nutrients in the bioengineered food differ from those in traditional varieties;
- the sources of the newly introduced genetic material come from a food plant associated with allergies;
- the food from the new variety differs significantly in composition from food of comparable varieties;
- the food contains marker genes that theoretically may reduce the therapeutic effects of clinically useful antibiotics;
- the plants are developed to make substances like pharmaceuticals or polymers, and will also be used for food; or
- the food to be used for animal feed has changes in nutrients or toxicants.

So far, most genetically modified foods have not required pre-market approval, although a few have had their composition changed significantly such that they were labeled differently. Most proteins from genes transferred into foods to give them new traits are either GRAS or otherwise exempt from regulation. A GRAS substance is excluded from the definition of a food additive.[15]

FDA has, however, instituted a voluntary consultation process whereby the developer can resolve any safety or regulatory issues prior to marketing. The agency has proposed keeping this process voluntary when it published a proposed rule in January 2001. The proposed rule would require that a food company notify the agency 120 days prior to marketing a bioengineered food and supply the agency with safety test data. FDA strongly urges companies to consult the agency prior to the mandated notification deadline in order to ensure agreement on the types of safety testing that will be needed. By the end of 2000, developers had consulted with FDA officials on 44 products: 12 times for corn; 5 for tomatoes; 8 for canola; 5 for cotton seed; 4 for potatoes; 3 for soybeans; 2 each for sugar beets and squash; and once for flax, radicchio, and papaya.

---

[15] Affirmation petitions for "generally-recognized-as-safe" (GRAS) status must be filed to gain FDA's formal agreement with a sponsor's independent determination that a substance is GRAS. FDA considers that bioengineered substances intentionally added to food are food additives. FDA must review and approve any food additives that do not have GRAS status before foods containing new additives can be introduced onto the market.

Case Study of a Commercial Genetically Modified Food: Flavr Savr Tomato

Although a few corporations have made substantial, long-term investments in the application of biotechnology to foods, not all genetically modified foods are commercially successful. Calgene's Flavr Savr tomato, for example, had limited technical success and did not meet commercial expectations. Calgene genetically modified a strain of tomato to reduce activity of a particular enzyme (polygalacturonase) that affects softening of outer tissue during ripening. Because the genetically modified tomato had less of this enzyme, it could remain longer on the vine prior to harvest thereby enhancing its tomato flavor. However, Calgene chose to manipulate genetically a tomato strain that had qualities more useful for processing than for the fresh market. The Flavr Savr never achieved commercial success because it cost more and did not taste better than competing cheaper tomatoes.

After its founding in 1980, Calgene conducted basic molecular biology research to investigate optimum techniques for the genetic manipulation of foods. In the mid 1980s, Campbell Soup co. decided to support Calgene's research efforts to produce a genetically modified tomato, presumably for use in Campbell's products. In 1989, officials at Calgene initiated discussions with FDA regarding the regulatory status of the tomato. FDA responded in 1992 by publishing its policy on foods derived from transgenic plants. Essentially, FDA decided to regulate genetically modified foods as any other food derived from traditional practices. The genetic technology needed to alter the tomato used an antibiotic marker that produced very small amounts of a non-tomato protein in Flavr Savr. According to FDA, the protein was viewed as a food additive since it changed the tomato's composition. Calgene officially petitioned FDA in January 1993 to allow the presence of this protein as a food additive. In May 1994, FDA approved Calgene's petition.

This approval opened the way for commercial marketing of the Flavr Savr tomato. However, in 1993 with significant public opposition to the genetically engineered tomato Campbell Soup Co. decided not to use genetically modified tomatoes in its products. Calgene then began efforts to market Flavr Savr as a freshmarket tomato rather than for use in processing. However, the tomato bruised easily and was less firm than expected. This characteristic caused production, transportation, and distribution problems. Competition from new tomato strains bred by traditional methods was an additional obstacle.

As reports of problems with commercialization of Flavr Savr grew, Calgene's financial condition weakened. In June 1995, Monsanto acquired a 49.9% equity stake in Calgene through the purchase of Calgene stock. In August 1996, Monsanto acquired controlling interest in Calgene. Monsanto emphasized other research programs at Calgene and subsequently moved control of Flavr Savr to another of its subsidiaries, Gargiulo Inc. in Naples, FL. According to industry sources, Gargiulo discontinued the effort to commercialize Flavr Savr.

In contrast, a British company, Zeneca, succeeded in marketing a genetically modified tomato in England. Now grown in California and processed into tomato paste for sale in the United Kingdom (U.K.), the Zeneca tomato has a label declaring that the paste is produced from genetically modified tomatoes. This product has since been withdrawn from the market because of consumer protests.

Development of this tomato used genetic technology virtually identical to that used by Calgene. Zeneca chose to genetically modify similar ripening-related enzymes in a tomato strain that had desirable processing characteristics. The Zeneca tomato yields a paste that is perceived as thicker and more flavorful than other tomato pastes. When combined with a comparatively low price plus marketing efforts through established market chains in England, this product has become the best selling tomato paste in the U.K.

Source: J.H. Maryanski. "FDA's Policy For Foods Developed By Biotechnology." *Genetically Modified Foods: Safety Issues* (Ed.s, Engel, Takeoka and Teranishi). American Chemical Society, Symposium Series 605, Ch. 2, 1995. p. 12-22; "Biotechnology of food." *FDA Backgrounder*. May 1994. [http://www.fda.gov/opacom/backgrounders/ biotech.html.]; M. Schechtman, Animal and Plant Health Inspection Service, USDA. "Assurance of Environmental Biosafety for Agricultural Products Derived through Modern Biotechnology: The Case Study of a Delayed Ripening Tomato." *The Organization for Economic Co-Operation and Development Environment Monograph No. 107*, Report of the OECD Workshop on The Commercialization of Agricultural Products Derived Through Modern Biotechnology. [www.oecd.org/ehs/mn107se.htm.]; M. Groves. "The Cutting Edge." *Los Angeles Times*, August 18, 1997; S. Lehrman. "Biotech Tomato Bruised." *San Francisco Examiner*, January 10, 1993.

On a food product's label, the FFDCA requires producers of foods to describe the product by its common name and reveal important facts associated with claims made or suggested. The agency interprets the FFDCA as not giving it authority to mandate labeling based solely on a consumer's "right to know" the method of production if the final product is considered "safe." Therefore, the agency does not mandate labeling to indicate the method by which a new variety was developed (e.g., that it was genetically engineered). However, the FFDCA does require that all information on labels be truthful and not misleading. Special labeling may be required if the developed food significantly differs from its conventional counterpart such that the common name would no longer apply. For example, FDA required the renaming of a soybean oil whose fatty acid composition had been altered by engineering. The new name, "high oleic acid soybeans," describes what is different about the oil but not its production method.

FDA requires special labeling for foods if they pose special safety or usage issues. For example, if a food had a new protein introduced into it to which people were allergic, FDA would require the label to reveal that information. In its 1992 policy statement, the agency noted that labeling would be required if genes were introduced from foods that were commonly allergenic, unless the developer could scientifically demonstrate that the protein was not responsible for the allergenicity of the original food. Examples of commonly allergenic food include milk, eggs, wheat, shellfish, tree nuts, and legumes. In one case, a developer demonstrated that DNA transferred to soybeans from a Brazil nut caused the production of a protein responsible for an allergic reaction. Consequently, the private developer discontinued research on that particular modified soybean. FDA has asked that developers of genetically modified food demonstrate that the introduced proteins do not share structural similarity to known allergens and are not resistant to digestive enzymes and acid.

On January 18, 2001, FDA published in the *Federal Register* a "Draft Guidance for Industry: Voluntary Labeling Indicating Whether Foods Have or Have Not Been Developed Using Bioengineering" and is seeking comments. In this document, FDA reaffirmed that it believes most genetically engineered foods are substantially equivalent to their conventional counterparts, and it decided it would not require special labeling of all bioengineered foods because it believes that the use of bioengineering, or its absence, does not itself cause a material difference in the food. However, the agency did suggest that because of the strongly divergent views on labeling, manufacturers may consider providing more information about bioengineered food. The information given, however, must be shown to be truthful and not misleading. To avoid false or misleading statements[16] about the absence of bioengineered ingredients (because there are no established threshold levels of bioengineered constituents or ingredients in foods), or to avoid implying that one food is superior to others, FDA suggests not using statements such as "GM free" or "biotech free." The agency does suggest the word "biotechnology" is preferred by some consumers over "genetic engineering" or "genetic modification." It also claims that if validated testing is available, it can be used to verify whether the label is truthful. Or manufacturers could keep records to document the reasons why a food's label is truthful. Comments on the draft guidance are requested.

That same day, January 18, 2001, FDA proposed to make mandatory a notification system whereby a food company would notify the agency 120 days prior to releasing a genetically engineered food onto the market.[17] Along with the notification must be data that show that the is safe. FDA is not proposing to require that genetically engineered foods undergo a "food additive" review — the extensive pre-market approval process. The agency is encouraging industry to consult FDA about the types of tests that could be needed to prove safety. FDA would then have time to review the company supplied data and would make it available to the public for comment. Proprietary information would not however be released to the public. Critics of this proposed rule think that independent testing of the safety data is necessary to restore consumer confidence in this food. They assert that the testing should be done either by the agency or independent contractors but not by the company that has a vested interest in showing only the product's safety.

U.S. Department of Agriculture (USDA), Animal and Plant Health Inspection Service (APHIS)

APHIS issues permits for the importation and domestic interstate shipment of certain plants, animals, and microbes that have the potential for creating pest problems in domestic agriculture. The agency has historically regulated pests that attack plants — any living stage of any insect, mite, nematode, slug, snail,

---

[16] U.S. Dept. of Health and Human Services. Food and Drug Administration. "Draft Guidance for Industry: Voluntary Labeling Indicating Whether Foods Have or Have Not Been Developed Using Bioengineering." *Federal Register*, v. 66, no. 12, January 18, 2001. p.

[17] U.S. Dept. of Health and Human Services. Food and Drug Administration. " Premarket Notice Concerning Bioengineered Foods"(Proposed Rule). *Federal Register*, v. 66, no. 12, January 18, 2001. p. 4706-4738, to amend 21 CFR Parts 192 and 592.

protozoa, and/or other invertebrate animal, bacteria, fungi, or parasitic plant or reproductive part. It is also interested in viruses — infectious substances that would directly or indirectly injure or cause disease or damage plants or plant parts or any processed, manufactured, or other plant products.[18]

For new plants that could become pests, APHIS issues site specific permits for field tests or for release into the environment. The agency reviews permit applications and prepares an environmental assessment in which it evaluates the probable environmental impact of the release. The permit application process requires that the developer disclose information about the development of the plant and that appropriate facilities and control measures are in place during transport and field tests. If the agency reaches a "Finding of No Significant Impact" (FONSI), a permit is issued.[19] Before decisions are made, APHIS seeks concurrence with states on regulatory actions.

In 1993, APHIS introduced an expedited procedure for approving limited permits so that field testing of six crops could begin without a completed formal application that included an environmental assessment. For genetically engineered plants that meet certain eligibility requirements and performance standards, the sponsoring company need only submit a "notification" letter to the agency, a modified and abbreviated application which describes the gene, where the tests will take place, and the characteristics of the plant. The agency has 30 days to process the application before the sponsor can proceed with the field test. In 1997, APHIS expanded the expedited procedure to cover many more crops; by 1998, 99% of all applicants used the expedited process.

After tests are completed and an application is submitted, APHIS has 120 days to decide whether the product poses a risk of being a plant pest and whether a product is ready for full "release" onto the market. APHIS then completes an environmental assessment before making its decision.

---

[18] 7 CFR 340, et seq.

[19] U.S. Congress. OTA. New Developments in Biotechnology — Field Testing Engineered Organisms. *Genetic and Ecological Issues*, no. 3, OTA-BA-350, May 1988. Washington, GPO, 1988. Chapter 3.

Case Study of a Genetically Modified Hormone: rBST

Bovine somatotropin (BST), also referred to as bovine growth hormone (BGH), is produced within the pituitary gland of all cows. It helps in the lactation process and is a normal trace constituent of milk. Farmers inject doses of the genetically modified hormone (called commercially Posilac) into cows to enhance milk yields and lengthen the lactation cycle. The result is an increase in milk yields of up to 15%. Prior to the 1980s, BST treatments were experimental and costly, since extractions of bovine tissues were the only source of the compound. With recombinant DNA technology, the supply of recombinant BST (rBST or rBGH) is more abundant and less expensive.

Critics are concerned that excessive amounts of rBST could compromise human and animal safety. They state that FDA did not provide adequate review of data to establish safety of the product prior to its approval. Six scientists within Health Canada, the Canadian federal agency involved in approving drugs, contend that the drug may not be safe for human consumption. However, the expert panel charged by Health Canada to review the drug found no risks to human safety. On January 14, 1999, Health Canada rejected approval of rBST based on concerns for animal health.

In June 1992, a joint expert committee of FAO/WHO concluded that rBST is safe for use and that Maximum Residue Limits (MRLs) are unnecessary. After a second review in February 1998, the committee arrived at the same conclusion. On March 10, 1999, the EU Scientific Committee on *Animal Health and Animal Welfare* recommended that the current EU ban on the use of rBST should continue in effect.

In 1989, Monsanto petitioned FDA's Center for Veterinary Medicine to approve an rBST product as a new animal drug. The review process took 4 years and was more extensive than most approval processes due to the controversial nature of the product. In August 1990, FDA published a review of data on rBST and concluded that it "presents no increased health risk." An August 1992 GAO report suggested that there may be an increase in mastitis in cows treated with rBST. The report also suggested there could possibly be indirect human health effects from residues of antibiotics used to treat cows for mastitis (udder infections). On March 31, 1993, an advisory committee of FDA's Center for Veterinary Medicine concluded that adequate safeguards are in place to prevent unsafe levels of antibiotic residues from entering the milk supply due to increased mastitis in rBST treated cows.

FDA approved the rBST product, Posilac, on November 5, 1993, with the stipulation that its developer, Monsanto, conduct a post approval monitoring program (PAMP) to provide further information related to possible effects on animal health experienced by rBST treated cows. FDA publishes occasional PAMP updates to summarize clinical manifestations associated with rBST treated cows. Since 1994, there have been 1,235 reports of adverse reactions in cows treated with Posilac although FDA states "the reported clinical manifestations are known to occur in dairy cattle not supplemented with Posilac." It also indicates that the number and types of reported effects raise no new animal health concerns.

On February 7, 1994, FDA offered interim guidance on labeling of milk from untreated cows, since some companies wanted to label their milk products as "BST-free." Products may be labeled as coming from animals not treated with rBST, but since BST is a normal constituent of milk, FDA determined that it is misleading to label milk as "BST-free." In May 1994, FDA's Food Advisory Committee and Veterinary Medicine Advisory Committee discussed whether foods derived from cows given supplemental rBST should be labeled as such. The committee report states "deliberations indicate that any method for instituting labeling for food from BST-supplemented cows would have to resolve many difficult scientific and policy questions." In 1999, FDA requires no labeling of milk products produced from cows supplemented with rBST.

On December 15, 1998, the non-profit Washington, D.C.-based Center for Food Safety petitioned FDA to withdraw approval of rBST citing possible health effects not addressed by FDA. As of January 2001 this issue is pending within the agency.

U.S. Department of Agriculture, Food Safety and Inspection Service (FSIS)

Developers of transgenic animals must submit data to FSIS to prove that the livestock and poultry involved in biotechnology experiments are not adulterated and can be slaughtered and sold as food with other beef and poultry. Prior to approving slaughter and sale, FSIS inspectors look at the number, age, sex, and other factors. The ultimate disposition of transgenic animal carcasses, whether by rendering, slaughter for food, or other disposal is of concern to regulators.[20]

Environmental Protection Agency

The EPA regulates pesticides. Pesticides are broadly defined as any substance or mixture of substances intended for "preventing, destroying, repelling, or mitigating" pests. EPA currently refers to "plant-pesticides," now called "plantincorporated protectants," as plants that produce pesticides within their tissues.[21] Scientists have also genetically engineered plants that are resistant to specific herbicides. Although herbicide resistant plants are not "plant-incorporated protectants," they are subject to EPA regulation since they can affect the use of herbicides.

EPA regulates plant-incorporated protectants, under the Federal Insecticide, Fungicide and Rodenticide Act (FIFRA) and FFDCA. Under FIFRA, EPA determines the risk the pesticide substance in the plant poses to humans and the environment and approves registration of those substances for particular uses that will not generally cause unreasonable adverse effects. This determination involves balancing risks from the pesticide with benefits associated with its use.[22] A pesticide (including a plant-incorporated protectant) cannot be sold or distributed in the United States unless it is registered with EPA.

---

[20] At the FDA Science Forum on December 9, 1998, Dr. Steve Sundlof of the Food and Drug Administration stated that appropriate disposition of transgenic animals is becoming a concern as their numbers increase. He recommended that companies needing to dispose of carcasses consult with Center for Veterinary Medicine.

[21] EPA has proposed new nomenclature in the *Federal Register,* "Plant-Pesticides, Proposed Rule, Supplemental Notice of Availability of Information,"April 23, 1999 (v. 64, no.78) to amend 40 CFR parts 152, 174 and 180.

[22] International Life Sciences Institute, Health and Environmental Sciences Institute. An Evaluation of Insect Resistance Management in Bt Field Corn: A Science-Based Framework for Risk Assessment and Risk Management. Washington, D.C. 1998.

If the plant producing the plant-incorporated protectant is a food crop, EPA must establish a "safe level" of pesticide residue allowed, a tolerance level, under the authority of Section 408 of the FFDCA. A "safe level" of the pesticide residue is defined as that level at which there is "a reasonable certainty that no harm will result from aggregate exposure to the pesticide chemical residue, including all anticipated dietary exposures and all other exposures for which there is reliable information."[23] Because no tests of the registered (approved) plant-incorporated protectant have shown toxicity to humans so far, EPA has given them an exemption from the requirement for a tolerance level.[24]

Table 2. EPA Registered Plant-Incorporated Protectant

Pesticide substance	Crop	Registrant	Year Registered
Bt Cry3A	Potato	Monsanto	1995
Bt Cry1Ab	Corn	Mycogen/Novart is	1995/8
Bt Cry1Ac	Cotton	Monsanto	1995
Bt Cry1Ab	Corn	Monsanto	1996
Bt Cry1Ab	Corn	Novartis	1996/8
Bt Cry 1Ac	Corn	Dekalb	1997
Bt Cry 9C	Corn	AgrEvo	1998
Potato Leaf Roll Virus	Potato	Monsanto	1998

Source: Environmental Protection Agency, Biopesticides and Pollution Prevention Division.

EPA has registered few plant-incorporated protectants. **Table 2** shows that EPA has registered three genetically modified crops containing plant-incorporated protectant: potatoes, cotton, and corn (including field corn, sweet corn, and popcorn). **Table 3** shows that EPA has exempted from the requirement for a tolerance several genetically engineered viral coat proteins that can be used in food commodities. So far, all but one EPA-approved products contain a "Bt" deltaendotoxin. The delta-endotoxins are proteins, one of the many toxins that may be naturally produced by the bacterium, *Bacillus thuringiensis,* and are species-specific, affecting only certain insects. They are also virtually harmless to humans and animals. When a susceptible insect consumes the protein, its digestion is severely disrupted, further feeding stops, and the pest eventually dies, usually within 2 days.

[23] Section 408 of the FFDCA as amended by the Food Quality Protection Act of 1996.

[24] 40 CFR 180.1155. The plant-incorporated protectant registered so far contain *Bacillus thuringiensis*, listed in **Table 2**, and have exemptions from the tolerance requirements.

Table 3. Exemptions of Viral Coat-Proteins[a] from Requirements of a Tolerance

Viral Coat Proteins
Watermelon Mosaic Virus-2 and Zucchini Yellow Mosaic Virus — in or on Asgrow line ZW20 of *Cucurbita pepo* L.
Potato Virus Y — in or on all food commodities
Potato Leaf Roll Virus — in or on all food commodities
Zucchini Yellow Mosaic Virus — in or on all food commodities
Watermelon Mosaic Virus-2 — in or on all food commodities
Papaya Ringspot Virus — in or on all food commodities
Cucumber Mosaic Virus — in or on all food commodities

Source: Environmental Protection Agency, Biopesticides and Pollution Prevention Division.

[a] Viral coat proteins are components of the outer cell wall that encloses a virus' genetic material. Expression of the coat protein in the plant confers resistance to the virus by a mechanism known as cross-protection.

In 1994, EPA proposed a rule to refine its regulatory oversight of plantincorporated protectant. As a part of the rule, EPA included in its definition both the plant-incorporated protectant and the genetic material transferred into the plant. Under this rule, EPA proposed to exempt several categories of plant-incorporated protectant from FIFRA and FFDCA section 408 requirements.[25] Under this rule, EPA proposed to exempt: (1) plant-incorporated protectants derived from plants sexually compatible with the recipient plant; (2) plant-incorporated protectants that act by primarily affecting the plant;[26] and (3) plant-incorporated protectants based on a coat protein from a plant virus. These rules were designed to exempt certain categories of substances that EPA believes are low risk based on familiarity and presence in the food supply, e.g., plant hormones and coat proteins from plant viruses. The agency believed that these proposed exemptions would result in little or no effects on plants produced by conventional plant breeding. They could, however, affect research into innovative plants, according to critics, because of the additional regulatory burden that the rule would represent. The rule was finalized on January 18, 2001.

---

[25] 40 CFR 152.20

[26] An example of a plant-pesticide acting to affect the plant would be the modification of hairs on tomato plants to inhibit aphids from feeding on the plant.

EPA'S Final Plant-Incorporated Protectant Regulation

In the January 18, 2001 final rule, EPA clarified three regulations governing plant-incorporated protectants. With this regulation, EPA intends to ensure that plant-incorporated protectants derived from genetically engineering will meet federal safety standards under both the FFDCA and FIFRA. EPA will also set food tolerances for residues of plant-incorporated protectants (or give exemptions) found in foods. The agency also exempted from registration plant-incorporated protectants derived from conventional breeding of sexually compatible plants and genetic material necessary for the production of plant-incorporated protectants. The final rule did not exempt plant-incorporated protectants derived through genetic engineering from sexually compatible plants; plant-incorporated protectants that act primarily by affecting the plant; and plant-incorporated protectants based on viral coat proteins. Both these exemptions were questioned in the National Academy of Sciences report, published in April 2000, entitled "Genetically Modified Pest-Protected Plants: Science and Regulation." EPA has asked for comments on the report's recommendations.

In September 2000, StarLink corn containing a Cry9C protein, a protein approved only for use in animal feed, turned up in taco shells. Without approval for human use, or exemption from approval, the Cry9C protein is considered an adulterant. The current regulatory system would still not necessarily have caught this adulterant in human food, even if there had been a mandatory notification and labeling program. The incident and subsequent problems with StarLink corn being found in shipments that would have been used in food, however, has led EPA to state publically that it will probably never again allow the entry of bioengineered products onto the market that cannot be used in human food. The incident also led FDA to develop sampling and testing guidance for the industry so that testing results could be used to verify the labeling of corn with or without the Cry9C protein.

ISSUES

Congress' attention will likely be drawn to a number of biotechnology issues due to public concern. Issues include concerns about public health, religious issues, labeling, the environment, the economic impact of this technology, and international trade competitiveness.

Labeling for Public Health and Religious Practices

Public Health Concerns

Many consumers express wariness of new "supercrops" and novelty foods, fearing that introduced genes could prove allergenic or harmful to human health. For example, if new genes inadvertently caused a plant to produce toxins at higher levels than are present naturally, there could be long-term health consequences for humans.[27] Some consumers are worried that a gene introduced into plants to protect against pests could also cause the plant to alter its pollen, thereby affecting the health of humans prone to some sensitivities.

Some critics are dismayed that FDA is placing all the responsibility on manufacturers to generate safety data, as it does normally under its pre-market approval system, and is reviewing only the conclusions of industry-sponsored studies, rather than conducting its own tests.[28] Critics have asked that more tests be conducted for chronic effects prior to products being placed on the market to ensure that all uncertainties regarding human health be explored.[29] Proponents argue that additional testing of genetically engineered foods is unnecessary because all foods must meet the same federal safety standards regardless of whether they are genetically engineered.

There is a growing movement among consumer groups that advocate the labeling of all genetically modified foods (GMO foods) that were produced through the process of genetic engineering. This position reflects a policy of "consumer sovereignty" or the "right to know," which supports the disclosure of

---

[27] In August 1998, Dr. Arpad Pusztai, a scientist from the United Kingdom, told a television team that a potato that he had genetically engineered to contain a lectin (a naturally occurring insect resistant protein) had led to harmful health effects in rats. (The 10-day feeding trials showed rats with weakened kidney, thymus, spleen, stomach wall, and immune system damage. The rats' brain size decreased.) His research was disowned by the institute for which he worked and he was forced to resign. On February 16, 1999, The Guardian reported that 21 scientists who reviewed the data from the experiments found it credible and protested this treatment. Michael Sean Gillard, Laurie Flynn, and Andy Rowell. "Food Scandal Exposed." The Guardian, February 12, 1999. p.1. Other scientists think that the quality of the potatoes fed to the rats may not have been consistently monitored and may itself have caused these problems ("Seeds of Discontent," The Economist, February 20, 1999, p. 75-77.)

[28] FDA typically uses data generated by individual companies to make regulatory decisions for food additives, drugs, medical devices, and biologics.

[29] Lehman, H. and J.F. Hurnik, "Concerns about the Ethics of Genetic Modification." Paper presented at the Fifth World Congress on Genetics Applied to Livestock Production, August 7-12, 1994. Proceedings, v. 20, as found at: [http://www.oac.uoguelph.ca/www/CRSC/faculty/eac/lehman.htm]

all relevant information on a label so consumers may make food choices based on their own values.[30]

Religious Practices

Others, particularly religious groups, are concerned that foods might contain genes from animals, such as swine prohibited by some religions, and they maintain that they have a right to know if foods contain those genes.[31] Both the kosher (Jewish) and halal (Muslim) communities have mechanisms in place to determine which products are acceptable to their adherents, and thus have not concerned themselves with secular labeling issues.[32] However, both Orthodox Rabbis and Muslim leaders have ruled that simple gene additions that lead to one or a few new components in a species are acceptable for kosher and halal law.[33] The Muslim community has not yet resolved whether a gene derived from swine is an exception to the above acceptance. For example, both groups have raised no objection to the use of bioengineered chymosin (rennin) in the production of cheese. The status of more significant changes in the genetic makeup of species remains to be decided. However, cloning, in particular, raises serious ethical/moral issues for religious leaders of all faiths. That discussion has involved a much broader range of clergy within the respective communities as well as other communities without dietary laws.

Labeling

Currently, no federal agency requires foods to be labeled as genetically modified. One reason is because the agency sees these products as substantiallly equivalent to traditional food products. This assumption has been supported by the courts. Under Vermont's 1995 mandatory rBST labeling rule retailers rather than producers paid the expense of enforcement to ensure that all milk produced using rBST was so labeled. A federal appeals court overturned this requirement, and since the spring of 1997, Vermont has authorized voluntary labels for rBST-

---

[30] Thompson, Paul B. "Food Biotechnology's Challenge to Cultural Integrity and Individual Consent." *Hastings Center Report,* v. 27, no. 4, July-August 1997. p. 34-38.

[31] Flamm, Eric. Office of Policy, Food and Drug Administration, conversation with authors, January 28 , 1999. Mr. Flamm also said that the arguments about religious freedom issues are theoretical so far because there are no products on the market that insert genes from animals into plant species.

[32] Regenstein, Joe, Professor of Food Science, Cornell University, conversation with the authors, February 12, 1999. CRS-19

[33] Chaudry, M.M. and J.M. Regenstein. "Implications of Biotechnology and Genetic Engineering for Kosher and Halal Foods." *Trends in Food Science Technology*, v.5, 1994. p. 165-168.

free dairy products.[34]

On January 18, 2001, FDA published a draft guidance for the industry giving examples of how the food industry could label foods that have or have not been developed using bioengineering. The guidance suggests that consumers prefer labels that explain why it was used in the food or how it is different from traditional food. The guidance suggests that all labels be "value-neutral." It also suggests that using the terms "GMO-free" or "not genetically modified" labeling could be misleading because such a label implies there is a zero level of bioengineered material which is almost always impossible to verify. The phrase "derived from biotechnology " would explain that the claim is being made and would allow consumers a choice about whether to purchase products produced by the new technology, to make better judgements about the compliance of the food product with ethical and religious beliefs, and lessen objections to its use.[35] Critics of this draft say that all labeling should be mandatory, not voluntary. They also claim that if the food companies are not forced to label genetically engineered foods, if there are health or safety problems, it would be more difficult to trace illnesses stemming from a particular bioengineered food.

The food industry generally opposes all biotech labeling because consumers may interpret these to be "warning labels," implying that the foods produced through biotechnology are less safe or nutritious than conventional foods. Food production interests believe that consumers, thinking that a product is different from conventional foods, may not gain the benefits from foods that have been modified genetically because they are uneasy with the technology and may not try the products. The industry supports FDA's reasoning in the draft guidance.[36]

Defenders of the agencies' policies suggest that to date genetically modified foods are similar to non-GMO foods and are easily regulated under the current

---

[34] Kolodinsky, Jane, Qingbin Wang, and David Conner. "rBST Labeling and Notification: Lessons from Vermont." *Choices*, third quarter 1998. p. 38-40. U.S. Court of Appeals for the Second Circuit. No. 876, August Term, 1995. (Argued: November 2, 1995 Decided: August 8, 1996) Docket No. 95-7819.

[35] Some have suggested that the industry use USDA's organic food labeling regulations, currently under development, as one way of telling consumers that foods are free of biotechnology products. Comment by Alan Goldhammer, scientific director of BIO, February 8, 1999.

[36] The agencies have authority under existing laws to regulate these new products in the same manner as those developed through traditional agricultural and food processing. Food companies have not been required to disclose their method of development on the label. For example, sweet corn does not need to be labeled "hybrid sweet corn" because it was developed through cross-hybridization. Nor did FDA require special labeling for the Flavr Savr tomato, because the new tomato is not significantly different from conventional tomatoes. If a new food contains a protein derived from a food that commonly causes allergic reactions, and the developer cannot demonstrate that the protein is not an allergen, labeling is mandatory to alert sensitive consumers, because they would not expect to be allergic to the food.

policy structure. If they are substantially different, these foods must be regulated as if they were food additives and receive FDA approval before being marketed. They state that current GMOs are not a risk as no scientific evidence shows an effect on human or animal health. For instance, most plant toxins are acute toxins and are well known, and tests have been developed to detect them quickly.[37] So defenders argue that there are no proven long-term health consequences for humans. However, few long-term studies have been completed.

On May 27, 1998, FDA was sued by a group of concerned citizens over the agency's "traditional" policy on the sale of genetically modified foods (*Alliance for Bio-Integrity v. Shalala*, D.D.C., No. 1:98CV01300, May 27, 1998). The group claimed that FDA's refusal to require labeling and safety testing raises health and environmental concerns and makes it difficult to comply with religious dietary laws. The suit identified 36 genetically modified foods being consumed daily without the knowledge of U.S. consumers. The suit cited both the First Amendment's protection of religious freedom and the Religious Freedom Restoration Act of 1993, which requires that federal laws or regulations not impede the free exercise of religion. The plaintiffs said that FDA's policy failed to abide by the public notice and comment procedures of the Administrative Procedures Act, and allowed genetically modified foods into the marketplace without being identified as such. The suit also claimed that FDA's policy is a burden to consumers' abilities to follow religious dietary laws. The lawsuit challenged FDA's policy that genetically modified foods are considered safe unless they contain substances identified in the 1992 policy which are allergens or would change the character of the food. The plaintiffs want the agency to carry out the same testing and safety evaluations conducted for food additive petition approval because, they argue, changes that might occur as a result of genetic engineering might include unwanted, unpredictable new toxins and/or carcinogens, elevated levels of inherent toxins and/or carcinogens, and/or degradation of nutritional quality. In particular, they wanted FDA to require the labeling of these foods because the foods have been changed "materially" and allegedly violate the FFDCA. The suit stated that "FDA is permitting unpredictable changes to the characteristics of certain foods which may be difficult for consumers to detect."[38] On August 7, 1998, FDA, through the Department of Justice, asked the U.S. District Court for the District of Columbia

[37] International Food Biotechnology Council, "Biotechnologies and Food: Assuring the Safety of Foods Produced by Genetic Modification," *Regulatory Toxicology and Pharmacology*, v. 12, 1990. p. S1- S196.

[38] Dern, Adrienne. "Justice Department Asks Court to Dismiss Lawsuit Challenging FDA's Policy on Genetically Engineered Foods," *Food Chemical News*, August 31, 1998. p. 10-13.

to dismiss this case because it felt that such charges are not unique to genetic engineering and the same changes occur in foods that have not been modified. On September 29, 2000, the Court accepted FDA's view that genetically engineered foods did not need special labeling because of consumer demand or because of the process used to develop these foods. Nor did the court require premarket approval of the food.[39]

Environmental Issues

Proponents of bioengineered crops claim that genetic modification can be less harsh on the environment than other technologies. They suggest that fewer agricultural chemicals might be needed to grow pesticide tolerant or insect resistant crops and that land would need less repeated tilling, which could lead to less erosion and soil infertility.[40] Supporters of genetic modification think that new developments contribute to environmentally sustainable development and greater food production. While they believe government regulatory efforts adequately ensure consumer and environmental health, they are aware of growing consumer concerns and some want to increase transparency of the regulatory and development process.

Critics have expressed strong concerns about the long-term risks and consequences of cross-pollination and of the disruption to the "cellular ecology" of plants. They state that the U.S. policy is based on the assumption of safety but there is little research on ecological or food safety risks. Scientists have shown that genetically modified rapeseed (canola) pollen was spread to wild radish weed relatives in nearby fields. The experiment demonstrated that it was possible to create new strains of weeds resistant to herbicides.[41] If such weeds emerged widely, farmers would need new, different, or stronger herbicides to counter their spread. Such super weeds could severely decrease crop productivity. Furthermore, some scientists have expressed concern that the widespread use of genetically engineered plants could alter the ecology of natural plant communities and of

[39] FDA Talk Paper: "U.S. District Court Dismisses Genetically Engineered Food Law Suit Against FDA." October 6, 2000.

[40] Doyle, John J. and Gabrielle J. Persley, eds. *Enabling the Safe Use of Biotechnology: Principles and Practice*, Environmentally Sustainable Development Studies and Monograph Series No. 10 (Washington, D.C.: The World Bank, 1996). p. 7. Also several biotechnology companies modified corn to contain Bt, to kill the corn borer and other pests. The corn borer causes huge losses. Corn with Bt had a 7% increase in yield per acre, bringing the farmer, on average, an increased net return per acre of $16.88. Moffat, Anne Simon. "Toting Up the Early Harvest of Transgenic Plants," *Science, v.* 282, December 18, 1998. p. 2176-2178.

[41] Brookes, Martin. "Running Wild," *New Scientist*, v. 160, no. 2158, October 31, 1998. p. 38-41.

wildlife food chains. Certain seed and herbicide companies agree with these critics, their point of view shaped by the possibility of development of "super weeds," rendering their products useless.[42]

Bt Resistance and Intervention Strategies

Concerns revolve around plants engineered to produce within their cells an insecticide called Bt that is produced naturally by strains of the bacterium *Bacillus thuringiensis*. The release of a study that showed in a laboratory that Bt corn pollen when eaten by Monarch butterfly larvae kill or stunt their growth engendered public concern.[43] Another concern is that large scale planting of crops containing Bt might lead to faster resistance development by insects. Critics of this technology state that large-scale production of engineered corn, soybeans and other foods will cause pests to develop resistance to Bt, thereby limiting its usefulness. It is not unusual for insects to develop resistance to pesticides that have been used for long periods of time. Organic farmers, in particular, are concerned because they do not have as many crop protection tools available as conventional farmers, and the loss of effectiveness of Bt could be a serious blow to their production. Due to these concerns, on February 18, 1999, Greenpeace, the Center for Food Safety, and some organic farmers sued EPA over its registrations of plant-incorporated protectant.[44] The suit calls on EPA to cancel all existing registrations for Bt crops, to cease the approval process for any new registrations, and to perform an environmental impact assessment analyzing the cumulative impacts from the current registrations.

Prior to these complaints, EPA determined that resistance to Bt can be slowed by requiring farmers to plant significant numbers of non-Bt seeds near the genetically modified resistant plants in designated separate areas called "refuges." The non-Bt plants allow pests to grow that will not develop Bt resistance. Since

---

[42] Critics also worry about altered organisms causing unpredictable and costly environmental damage over extended eco-ranges for sustained periods, and point to the cases of Dutch Elm disease, myxomatosis, and gypsy moths to illustrate their point of organisms spreading uncontrollably . Supporters claim that these invaders were not disease-causing GMOs, but were unpredicted consequences of unplanned introductions into new environments. Given multiple ecological problems related to farming practices such as nutrient runoff, for which there are no liability claims, it seems inappropriate to supporters that GMOs are lumped into these arguments. Critics suggest, however, that because genetically modified plantincorporated protectant may be unsafe, their release into the environment is irreversible and unpredicted consequences may be hard to mediate. U.S. Congress, Office of Technology Assessment, *New Developments in Biotechnology — Field Testing Engineered Organisms, Genetic and Ecological Issues*, no. 3, OTA-BA-350 (Washington, DC, May 1988) chapter 5.

[43] Kaesuk Yoon, Carol. "Altered Corn May Imperil Butterfly, Researchers Say," *New York Times*, May 20, 1999. p. A1 and A20.

[44] See website: [http://www.icta.org/legal/bt2.htm]

these pests will interbreed with their counterparts that eat Bt plants, there is less likelihood that a resistant super-pest could develop. If the "refuge" system is used by farmers, EPA estimates that the eventual development of resistance to Bt by insect pests could be extended 10 or more years.[45]

EPA now requires the use of insect resistance management (IRM) plans for some crops. EPA originally granted all the plant-incorporated protectants conditional registrations. With the registrations for corn and cotton to expire in 2001, as a condition of re-registration, companies will be required to develop effective IRM plans for these crops. Most re-registration applications have been submitted, and EPA is evaluating the information and could impose new conditions. Critics of the technology want mandatory IRM plans, for they claim that if the plans are voluntary, farmers will either not plant the nonresistant refuges or plant refuges that are too small to be effective.[46]

There is a voluntary IRM plan in place for potatoes, as recommended by a 1995 scientific advisory panel, and Monsanto is requiring compliance by growers who contract with it for the seeds.[47] Ongoing evaluation of IRM may result in additional mandatory requirements.

There is agreement, however, that toxic anti-pest substances produced by plants used for food must be proven safe. EPA points to an example of potato leaves that naturally contain a pesticidal substance that could cause birth defects. Since humans do not eat potato leaves, there is no need to regulate the substance;

[45] Environmental Protection Agency. Scientific Advisory Panel on *Bacillus thuringiensis* (Bt) Plant-incorporated protectant. Transmittal of the final report of the FIFRA Scientific Advisory Panel on *Bacillus thuringiensis* (Bt) Plant-incorporated protectant and Resistance Management. Meeting held on February 9-10, 1998. (Docket Number: OPPTS-00231).

[46] As part of the ongoing discussions with EPA and USDA, major see developers (accounting for more than 90% of the U.S. Bt seed corn market) have proposed that by the year 2000, they will require growers in contracts to plant 20% to 50% of their acreage in non-genetically engineered varieties. The agreement among the seed companies calls for a 20% refuge requirement in primary cotton-growing areas, with the minimum rising to 50% where Bt corn is grown in proximity to Bt cotton. "Industry Plan for Bt Resistance Management Gets Mixed Reaction," *Food Regulation Weekly*, v, 18, January 1999. p. 3-5. EPA is considering this proposal among its other options for implementation of IRM requirements. In the spring of 1999, EPA and USDA are planning a joint workshop to examine existing Bt crop IRM, including methods, implementation, grower's incentives, remedial actions if insect resistance is discovered, and compliance and enforcement issues. They are also interested in supporting the development and implementation of six to ten USDA Regional Pest Management Centers.

[47] Website: [http://www.monsanto.com/ag/asp/Monsanto.asp?MKTID=9999&PDTID=0] contains a statement, "We require every grower who decides to plant YieldGard corn to sign an agreement to establish a refuge adjacent to his or her YieldGard acreage to ensure a population of susceptible corn borers to mate with any naturally resistant borers that survive exposure to the Bt."

however, if spinach were genetically modified to contain the potato-leaf toxin, regulation would be needed.

Liability

There is also concern about liability. Who would pay if other crops or fields were ruined because of cross-pollination with these new seeds? An opponent of biotechnology, Jeremy Rifkin, says, "The insurance industry has quietly let it be known that it would not insure the release of genetically engineered organisms into the environment against the possibility of widespread environmental damage, because the industry lacks a risk-assessment science — a predictive ecology — with which to judge the risk of any given introduction."[48] According to Dr. L. Val Giddings, Vice President for Food and Agriculture, Biotechnology Industry Organization, the industry representatives directly dispute this claim, and say there is insurance available.

Economic Concerns

Critics are concerned about the growing presence of large agricultural conglomerate companies controlling the supply of seeds containing bioengineered traits. Growers' profits from bioengineered crops more than tripled from 1996 to 1997.[49] In 1996, U.S. farmers had planted mostly Bt cotton, Bt corn, and herbicide tolerant soybeans. In 1997, they expanded acreage for those crops and added herbicide tolerant cotton and potatoes.[50] Planted acreage has continue to increase for all crops except corn. Adoption rates vary by year, crop product, and location. They depend on the level of infestation of a targeted pest. They also depend on world commodity prices and the kind of corn that importers are willing to accept. Supporters of this technology cite convenience and lower costs as the main reasons for high farmer adoption rates.[51] But critics complain about the consolidation of patent ownership of this new technology which they believe constitutes monopolistic power in the marketplace. Their distrust is based in part

---

[48] Rifkin, Jeremy. "Genesis II; Commercial Prospects of Genetic Engineering and Biotechnology," *Across the Board*, v. 35, no.6, June, 1998. p. 29.

[49] Beachy, Roger N. Statement Presented to the House Committee on Agriculture, Subcommittee on Risk Management, Research and Specialty Crops, on behalf of the Council for Agricultural Science and Technology, March 3, 1999.

[50] James, Clive. "Global Review of Commercialized Transgenic Crops: 1998," *International Service for the Acquisition of Agri-Biotech Applications Briefs*, no. 8, ISAAA: Ithaca, NY; in 1997, profits totaled $315 million, up from $92 million in 1996.

[51] Riley, Peter A. and Linwood Hoffman. "U.S. Farmers Are Rapidly Adopting Biotech Crops," *Agriculture Outlook*, August 1998. p. 21-24.

on concern about the dependency such concentration brings to the agricultural sector and the possible abuse from such control.

The reasons for this concern vary. The concentration of control of patents in agricultural biotechnology raises concerns for some farmers and consumers that multinational corporations want to simplify the regulatory processes of governments to gain easier approval for their products without adequate review. Because so much of the technology is held by private companies, some regulators and researchers are concerned about how difficult it can be to obtain necessary information for appropriate regulation.

Others are concerned that large companies can conduct field experiments, sell seeds to farmers, and market genetically engineered products without appropriate attention to tests for the safety of consumers and the environment. Critics want more testing before farmers expand their planted acreage. There is uncertainty about how much testing is required, whether it should focus on the new genes themselves or on the substances they produce in the plant. Federal agencies can ensure adequate safety testing of biotechnology products by requiring companies to review impacts or by conducting tests for themselves. The FDA proposed rule, discussed earlier, could mandate that industry submit this data. However, others feel that independent testing is necessary to validate the safety of the food product.

There also appears to be concern about how much control companies wield over research in food and agricultural biotechnology. Companies hold patents and own specific germplasm[52] and research techniques needed for plant research. Some companies claim partial ownership of a food product created using their patented technologies. Biotechnology researchers have raised concerns that some private company scientists are not permitted to share innovations in research.

Most food biotechnology research is financed with private, not public, money, and the total amount spent is confidential. One newspaper reported that Monsanto estimates that research and development time and costs to create a commercial product are about 10 years and about $300 million. For every genetically engineered seed that goes to field trials, 10,000 have failed along the way.[53] Corporations charge steep prices for this technology, claiming the need to recoup their investment to be able to research the third and fourth wave of products.

---

[52] Germplasm refers to the basic genetic material of a species. There are germplasm "banks," for a variety of different life forms, where representative species are stored so that they can be reproduced for future studies that require their genetic material.

[53] Weiss, Rick. "Seeds of Discord," *Washington Post*, February 3, 1999. p. A1 and A6.

Most identifiable public funding for food biotechnology research for plant genome research is being sponsored by the U.S. Department of Agriculture (USDA). The Clinton Administration did not track federal food and agricultural biotechnology funding of research as a line item in federal budget analyses. Consequently, the total amount of public funding for this research is unclear.[54]

Public research funding under the authority of the Bayh-Dole Act of 1985[55] allows universities receiving grants for plant genomics research to hold the intellectual property rights for any useful discoveries. The Act has accelerated the linkages between university research and the creation of consumer products and contributed to the international competitiveness of U.S. industries. It has also encouraged research on minor-use crops.

International Trade Issues

At the present time, the United States is leading the world in biotechnological research, development of genetically modified organisms (GMO's as they are called internationally), and sales of the technology worldwide. The United States has no immediate challengers to this trading position. Some trade experts suggest that trading partners whose policies strongly reflect consumer concerns about the new technology are merely attempting to allow their own domestic industry time to develop a competitive position in this trade. For example, the European Union (EU) has now required that all GMOs be labeled as such, and has severely limited imports of food containing genetically engineered crops and foods. The United States, however, claims that there is not scientific basis to presuppose that genetically modified food products are more risky or substantially different from other products. U.S. officials believe that decisions on trade should be science-based and U.S. regulatory policy reflects this thinking. Such competition for leadership in biotechnology has influenced trade relationships among U.S. trading partners. Japan has diminished its import of U.S. corn after finding StarLink in their imports.

---

[54] Gabriel, Cliff, Deputy to the Associate Director of the Science Division, Office of Science and Technology Policy. *The White House, conversations with the authors*. February 23, 1999.

[55] Bayh-Dole Act, P.L. 96-517, §6(a), December 12, 1980, 94 Stat. 3019 (35 §§200 to 212); P.L. 98-620, Title V,§501, November 8, 1984, 98 Stat. 3364 (35§§201 to 203, 206 to 208, 210, 212); P.L. 99-502, §9(c), October 20, 1986, 100 Stat. 1796 (35 § 210).

Concluding Remarks

Recombinant DNA technology is producing revolutionary changes in agriculture. Supporters of this technology emphasize the potential benefits from these changes including the promise of higher yields and nutritionally enhanced foods from genetically modified commodities with reduced environmental impact. Opponents of the use of rDNA technology in agriculture are concerned about possible hazards to human, animal, and environmental health. They advocate more safety testing and labeling of products. Many are concerned about the possible consolidation and control of agricultural biotechnology by a handful of multinational corporations.

Since the technology potentially can provide greater yields, farmers are rapidly planting genetically modified seeds. As the U.S. experience with this new technology grows, it may become necessary to amend regulatory policies to protect public safety and the environment while allowing genetically modified crop development to progress. The agencies involved in regulating genetically modified foods (FDA, APHIS, and EPA) are implementing policies based upon a 1986 framework document that coordinates their regulatory activities for biotechnology products. This framework applies the same set of regulations to all food products and does not differentiate between foods that are produced with rDNA technologies and those that are produced by traditional methods. However, with the January 18, 2001 proposed rule, making mandatory the current voluntary consultation process whereby the industry now must formally notify the agency 120 days before releasing a food product onto the market. The proposed rule also lists the safety test data that must accompany the notification.

Although the United States is leading the world in the production and sales of genetically engineered products, a few foreign countries have resisted or diminished their U.S. commodity imports because of concerns over the safety of these bioengineered crops. Some suggest that foreign countries' resistance to genetically engineered crops can be traced to their desire to allow their domestic industry time to develop a competitive position in this trade and the growing unease in international trade relationships over the fast adoption by U.S. farmers of bioengineered crops. Others argue that Congress could exercise more oversight over the regulation of these food crops, fund more public research, and encourage the Administration to negotiate the easing of trade barriers and harmonizing standards.

Chapter 2

U.S. EUROPEAN AGRICULTURAL TRADE: FOOD SAFETY AND BIOTECHNOLOGY ISSUES[•]

Charles E. Hanrahan

INTRODUCTION

Differences over food safety measures and biotechnology have affected U.S.-European Union (EU) agricultural trade. Particularly contentious in both bilateral trade relations and in World Trade Organization (WTO) dispute settlement has been the EU's ban on meat produced using growth-promoting hormones. EU policy on bio-engineered products has also been a contentious issue although no formal disputes have been raised in the WTO. Developments in regulation and labeling of bio-engineered products could potentially affect future U.S.-EU agricultural trade.

Exports of $6.4 billion of agricultural products make the EU the United States' third largest market for U.S. farm products. Soybeans, which benefit from a zero-tariff binding and from an EU-wide ban on the use of meat-and-bone meal in livestock feeding, are the leading export. However, U.S. soybeans exports to the EU have declined–from $2.1 billion in 1996 to $1.1 billion in 2000-- as consumer concerns about bio-engineered crops have grown and the EU has sought to import non-biotech varieties from other sources. The EU also is the leading

[•] Excerpted from CRS Report 98-861.

U.S. export market for corn by-products (corn gluten feed), almonds, dried fruit, wine and tobacco. The EU had an agricultural trade surplus of around $1.7 billion with the United States in 2000. Although it remains a key market for U.S. farm products, the EU's share of total U.S. farm exports has declined from just over 30% in 1982 to around 15% currently.

The application of import restrictions in agricultural trade based on food safety considerations, such as the EU's meat hormone ban, is governed by the WTO Agreement on Sanitary and Phytosanitary (SPS) measures. That Agreement requires that SPS measures be applied only to the extent necessary to protect health and that they be based on scientific principles and on assessment of risk. The Agreement encourages countries to base their SPS measures on international standards and to recognize each others' standards (equivalency) as long as they achieve the same degree of protection. Disputes between WTO members over the requirements of the SPS agreement fall under WTO dispute settlement procedures, which were strengthened as part of the 1994 Uruguay Round multilateral trade agreements to make procedures more expeditious and decisions more binding. The ban effectively has eliminated U.S. beef exports to the EU, except for around 6,500 metric tons of non-hormone treated beef imported by the EU as part of an 11,500 ton import quota.

Differences over measures affecting trade in bio-engineered crops and foods also are an issue in U.S.-EU agricultural trade. Issues have been dealt with in bilateral U.S-EU discussions and, with the exception of the Biosafety Protocol Negotiations, in nonnegotiating fora such as the Organization for Economic Cooperation and Development (OECD) or the Codex Alimentarius Commission.[1] Bringing disputes over bio-engineered foods into the WTO appears problematic because there are no specific WTO rules and disciplines for bio-engineered

[1] *Biosafety Protocol for Genetically Modified Organisms: Overview* by Alejandro E. Segarra and Susan R. Fletcher, CRS Report RL30594, July 5, 2000. The OECD is an organization that provides for intergovernmental discussions among 29 industrial countries in the fields of economic and social policy. The Codex Alimentarius Commission is a joint activity of the U.N. Food and Agricultural Organization and the World Health Organization. The Commission drafts nonbinding standards for food additives, veterinary drugs, pesticide residues and other substances that affect consumer safety.

products in international trade.[2] Establishng such rules and disciplines may be on the agenda of the on-going WTO multilateral agricultural trade negotiations.[3]

THE MEAT HORMONE DISPUTE[4]

Since 1989, the EU has banned imports of meats produced with growth-promoting hormones. The international counterpart of internal EU regulations that prohibit use of hormones in domestic meat production, the ban has been justified as needed to ensure food safety and consumer health. Although entirely unrelated to hormone use, concern about "mad cow disease" or BSE (bovine spongiform encephalopathy), a disease of cattle shown to be transmissible to humans as Creudzfeldt-Jakob Disease (CJD), has created a climate in Europe that is unfavorable to resolving the meat hormone issue.

In 1996-97, the United States successfully challenged the ban as a violation of the SPS Agreement in WTO dispute settlement. WTO dispute and appellate panels ruled (August 1997) that the ban was not based on scientific evidence nor on an assessment of the risk to health from meat treated with hormones. The appellate panel left open the option to the EU to conduct a risk assessment of hormone-treated meat and a WTO arbitration panel subsequently ruled that the EU had 15 months from the date of the appellate decision to bring its ban into compliance with SPS rules.

Although the United States called for the immediate lifting of the ban, the EU kept it in place while it conducted its risk assessment. On May 13, 1999, the deadline for EU compliance with the WTO dispute ruling, the EU indicated that the ban would continue in force. As justification for continuing the ban, the EU Commission offered what it said was evidence that one of the U.S.-approved hormones is carcinogenic. U.S. trade and veterinary officials rejected the EU evidence which, they said, ignored and contradicted numerous scientific studies, including some by European scientists, that show "absolutely no human risks

---

[2] Trade lawyers differ over the need for *sui generis* rules and disciplines for bioengineered products in international trade versus other approaches such as interpreting or clarifying existing agreements to take them into account. With respect to the latter approach, if restrictions on bio-engineered products were justified on the basis of food safety concerns, then the provisions of the WTO's SPS Agreement could apply. Labeling could fall under the terms of the Technical Barriers to Trade (TBT) Agreement.

[3] The U.S. proposal in the current round of WTO agricultural trade negotiations calls for procedures for approval and regulation of the products of agricultural biotechnology that are transparent, predictable, and timely.

[4] For detail, including historical background, on this dispute see *the European Union's Ban on Hormone-Treated Meat*, by Charles E. Hanrahan, CRS Report RS20142.

associated with consumption of beef from animals treated with growth-promoting hormones." With WTO approval, the United States retaliated against the EU by imposing prohibitive tariffs on $116.8 million of EU agricultural imports.

The EU has offered to negotiate compensation in lieu of retaliation, but the United States has held to its position that compensation would be acceptable only as an interim solution until the EU lifted the ban. The EU's approach to compensation focuses on increasing the quota for non-hormone treated beef and reducing the 20% in-quota tariff. Under a compensation scheme, U.S. retaliation would be reduced by the amount of the increased value of exports of non-hormone treated meat. The EU seeks sanctions relief especially for French exports, particularly hard hit by U.S. retaliation.

U.S. beef producers, initially receptive to an enlarged quota, became disenchanted with the compensation negotiations when (May 24, 2000) the EU Commission announced that it was banning outright one of the hormones and that the ban on the others would be maintained provisionally. U.S. meat industry representatives expressed concerns that such an approach meant that the EU would in the future ask for WTO approval to remove the U.S. retaliation (and thereby end the compensation) because the ban would then accord with WTO rules.[5] The prospect of ending compensation without lifting the ban, according to the meat industry representatives, constitutes a disincentive for the industry to invest in the additional production of non-hormone treated beef needed to fill an expanded quota.

Resolution of the hormone dispute could remove a critical irritant to the overall U.S.-EU trade relationship.[6] Trade experts indicate that the meat hormone dispute has spilled over into other trade disputes such as the EU's successful challenge to U.S. tax breaks for off-shore export companies.[7] How the hormone issue is resolved could have important implications for future disputes over SPS measures. The WTO meat hormone decision is a strong affirmation of the Uruguay Round SPS Agreement and its requirements that countries base their SPS measures on scientific justification and risk assessment. At the same time, the dispute illustrates the difficulty of enforcing multilateral rules and procedures that are in conflict with popular concerns and national political decisions.[8]

---

[5] "Industry Letter on U.S.-EU Beef Talks", *Inside U.S.* Trade, November 10, 2000.

[6] Trade Conflict and the U.S.-European Union Economic Relationship by Raymond J. Ahearn, CRS Report RL30732, November 8, 2000.

[7] See The Foreign Sales Corporation (FSC) Tax Benefit for Exporting and the WTO by David Lee Brumbaugh, CRS Report RS205171.

[8] Marsha A. Echols, "Food Safety Regulation in the European Union and the United States: Different Cultures, Different Laws," *Columbia Journal of European Law*, Columbia University, Summer 1998, pp. 525-543.

BIO-ENGINEERED PRODUCTS

Bio-engineered crops and food products that contain them are leading to conflicts that have already disrupted U.S.-European trade in soybeans and corn. At issue are both approval procedures and labeling requirements for genetically modified (GM) products (crops and processed foods). Underlying the disputes are pronounced U.S.-EU differences over environmental effects of GMOs and the safety of GM foods.

U.S. farmers have rapidly adopted GM crops. More than 20 bio-engineered crops are now sold commercially and others are being developed. Among those adopted are herbicide-, insect-, and disease-resistant hybrids and varieties, mainly of soybeans, corn and cotton. In 2000, an estimated 54% of soybean acreage and 25% of corn acreage (down from 33% in 1999) in the United States are planted with GM varieties.[9] Reduced input costs and enhanced production flexibility (e.g., reduced tillage) are among reasons cited by farmers for adoption of GM varieties. Europe has been slower to accept and introduce GM crops, although the European Union has approved some GM varieties of soybeans and corn--Monsanto's Roundup Ready soybeans and Novartis's BT corn. (Monsanto's soybean variety is resistant to the herbicide Roundup. BT corn contains a gene that gives the plant resistance to the corn borer.)

Although some consumer, environmental, and scientific groups in the United States are calling for more regulation and monitoring of GM crops and food products, polls and surveys indicate that consumers in the United States have been largely accepting of bioengineered products and their safety for human consumption. In the EU, however, there has been widespread concern among consumers about the quality and safety of the foods they consume, including GM foods. Official handling of the BSE ("mad cow disease") crisis is often cited as having undermined consumers' confidence in food regulatory agencies and scientific assurances about food safety. The discovery of dioxin in poultry feed in Belgium and recent outbreaks of "mad cow disease" in EU countries thought to be BSE-free have exacerbated food safety worries. Environmental groups, such as Greenpeace, have been particularly active in opposing the introduction of GM crops and consumption of GM foods.[10] Views espoused by Greenpeace are widely shared by organized consumer groups throughout the EU. These groups and some scientists in the EU maintain that the long-term effects of GM products on health

[9] U.S. Department of Agriculture, National Agricultural Statistics Service, *Acreage*, June 2000.
[10] Position taken by Greenpeace can be found at [http://www.greenpeace.org/~geneng/]

and the environment are unknown and that GM products should bear cautionary labels to inform consumers about their contents.

The U.S. approval process for GM crops has facilitated their introduction into U.S. agriculture and their use in food processing.[11] The U.S. position has been that GM foods are no different from non-GM foods. Thus existing regulations for approving them are appropriate and adequate. Labels are not required, with some exceptions, although voluntary labeling as to GM content is permitted. The EU, however, maintains a separate regulatory system for GM products; requires mandatory labeling; and currently has in place a de facto moratorium on the approval of new GM crops while the European Commission develops new EU-wide legislation for approving and regulating bioengineered products.

The EU Commission has been working on a new regulatory framework for GM products for some time. The Commission's role is not only to decide on the introduction of GM crops into EU agriculture, but also to regulate their use. Even after Commission approvals there can be delays in actually introducing the crops into EU agriculture. For example, France, which had initially requested the approval for BT corn, subsequently prohibited its sale. Austria and Luxembourg continue to prohibit the use of GM crops. Approval by the EU Council of Ministers of three varieties of GM corn after months of delay has been blocked by France, causing the loss, according to the U.S. Department of Agriculture (USDA), of an estimated $220 million annually of corn exports to Spain and Portugal.

As to labels, the EU argues that they should be mandatory so as to respond to consumer desires for information about the food products they consume. Food processors in the EU are required to perform mandatory DNA or other laboratory tests to determine the GM content of products and label them accordingly. The United States, on the other hand, has opposed mandatory labeling, with some exception on the basis that available scientific evidence shows that food products made from GM crops approved for human consumption pose no threat to human, animal or plant health, and therefore should not need special labels.

Some U.S. companies, like their EU counterparts, have been taking steps to respond to perceived consumer demand, both at home and abroad, for GM-free foods or approved GM products. A number of U.S. food companies, including Heinz, Gerber, Frito-Lay and McDonalds have announced that they will not use GM ingredients in their products. The recent discovery of the presence of a GM corn, the StarLink variety, approved for animal feeding, but not human

[11] Food Biotechnology in the United States: Science, Regulation, and Issues by Donna U. Vogt and Mickey Parrish, CRS Report RL30198, June 2, 1999.

consumption, in tacos and some other products also evoked a response from the grain processing sector.[12] A.E. Staley, the U.S. corn-processing unit of a British commodity firm, Tate and Lyle PLC, has suggested that farmers consider raising only non-GM crops. Archer, Daniels, Midland (ADM), a processor of corn and soybeans, has urged its suppliers to produce only crops that have "full feed and food approval world-wide."[13]

Recent Regulatory Developments

Recent developments in both the United States and the EU indicate that their respective regulatory frameworks for GM foods are evolving. The EU appears ready to eliminate its de facto moratorium on approvals of GM crops under an agreement reached in early December 2000.[14] According to reports, the moratorium could be lifted as early as February 2001. Under a revised approval system, the European Commission will submit applications to plant GM crops individually to a committee of experts. The Commission would then seek member governments collective approval for products endorsed as safe by the experts provided the companies requesting approval agree to be bound by the legal framework to be established under the December 12 agreement. The legal framework would require that licenses for GM products be renewed periodically, that the public be consulted more widely, and that companies provide clear labeling.

The EU is also moving toward the establishment of an EU-wide food safety agency with advisory responsibilities, whose charge would be to assure EU consumers of the safety of foods they consume, including foods with GM ingredients. According to the EU's Commissioner for Health, David Byrne, such an authority would, among other things, restore citizens' confidence in biotechnology for agriculture and food through better information and communication. The new food authority would also, he said, bring a coherent and predictable regulatory framework to GM foods, feeds, and seeds.[15]

---

[12] *StarLink Corn Controversy: Background*, by Alejandro E. Segarra and Jean M. Rawson, CRS Report RS20732, November 17, 2000.

[13] "Companies: Grain Processors Apply New Pressures to Biotech Industry–U.S. Farmers Get Warnings Against Modification of Crops," *Wall Street Journal*, European Edition, November 20, 2000, p. 6.

[14] Deborah Hargreaves, "Deal to Revise Laws on GMOs", *Financial Times*, London: December 13, 1000,. P. 2; and "New GMO Laws Imminent Say EU Officials", *AgraEurope,* AE1931, December 15, 2000.

[15] "EU to Continue Efforts for Consumer Acceptance of G MO Products," *SCI Policy Report*, June 26, 2000, p. 4.

In the United States, the Clinton Administration (in May 2000) announced plans to require biotech companies to notify the Food and Drug Administration (FDA) four months before marketing a GM food. FDA would also create a regulatory mechanism by which foods could be labeled, voluntarily, as free of GM ingredients. USDA would also become directly involved in validating new scientific tests to detect the presence of GM ingredients in foods. More recently, in response to the StarLink controversy, USDA has begun the process of developing standards for separating biotech crops from conventional ones in calling for public comment on the issue by February 28, 2000.[16]

The U.S.-EU Biotechnology Consultative Forum, a panel of experts, selected to advise the United States and the EU on the pros and cons of biotechnology, including GM crops and GM foods in June 2000 made its report to the U.S.-EU summit on December 18, 2000 in Washington. Among its recommendations, was a proposal that the United States and the EU establish "content-based mandatory labeling requirements for finished products containing genetic material."[17] What impact this recommendation would have as the new Administration takes its place in the Trans-Atlantic dialogue on biotechnology and other issues remains to be seen.

[16] For details, see [http://www.usda.gov/gipsa/rulemaking/current/fed.reg.htm]

[17] The EU-U.S. Biotechnology Consultative Forum: Final Report, December 2000, p. 16.

Chapter 3

STARLINK™ CORN CONTROVERSY: BACKGROUND[†]

Alejandro E. Segarra and Jean M. Rawson

INTRODUCTION

StarLink™ is a corn variety that has been genetically modified to contain an insecticidal protein derived from a naturally occurring bacterium (Bacillus thuringiensis, or Bt.) The Environmental Protection Agency (EPA) approved the gene-spliced variety of yellow corn in 1998 for use only as animal feed and set a zero-tolerance level for its use in human food based on the fact that this particular Bt protein does not break down easily in the human digestive system, is heat resistant, and could prove allergenic.

StarLink corn was detected in taco shells in did-September 2000. The StarLink variety constitutes between 0.4% to 0.5% of total U.S. corn production; however, a larger (and unknown) amount of corn currently in market channels may be commingled with StarLink corn. EPA is examining a request from Aventis, the manufacturer, to grant StarLink a temporary emergency exemption from the zero-tolerance standard. Japan, which imports 30% of total U.S. corn exports and does not permit StarLink to be imported for any use, has asked government officials to make sure that no incoming shipments contain StarLink-

[†] Excerpted from CRS Report RS20732.

commingled corn. Several bills were introduced in the 106th Congress to require labeling of foods containing genetically modified organisms (GMOs).

In addition, a bill was introduced at the end of the 106th Congress (S.3184) to amend the Federal Food, Drug and Cosmetic Act to require pre-market consultation and approval for foods containing GMOs.

The presence of StarLink™ corn in food has become the first test case of contamination of the food supply by a genetically modified organism (GMO). Among the issues this raises for Congress are: What steps might help alleviate the immediate problems for farmers, grain elevators, exporters and trading partners? Are further changes in the current statutory or regulatory framework needed to address the food and environmental safety issues related to agricultural biotechnology? Can regulatory policies be changed in such a way that support for innovation in crop and food technologies is not undermined?

This chapter examines the events that led to the current situation and provides an overview of its impacts.

WHAT IS STARLINK™ CORN?

StarLink is a trademark for several genetically modified corn hybrids produced by Aventis Crop Science of Research Triangle Park, N.C. (a German-French life sciences consortium) and distributed through several seed companies. StarLink hybrids contain a plant pesticide protein (Cry9C) derived from a common soil microbe (Bacillus thuringiensis, or Bt), which kills certain destructive pests of corn such as the European corn borer. StarLink also is one of a handful of the currently approved genetically modified (GM) crop varieties that contains "stacked genes" (i.e., more than one commercially desirable transgenic trait at once.) StarLink contains: (1) the insecticidal Bt Cry9C protein; and (2) genes from the bacteria Streptomyces hygroscopicus, which makes StarLink tolerant to a commonly used broad-spectrum herbicide.

StarLink & EPA: Regulatory Timeline

March 14, 1997	EPA issues an experimental use permit (EUP) to Plant Genetic Systems (PGS) to test corn seeds containing the Cry9C protein in 3,305 acres in 28 states.
August 8, 1997	EPA announces an application from PGS to register the Cry9C pesticide under FIFRA.
September 19, 1997	EPA announces that PGS asked for a full exemption of a tolerance for Cry9C residues in or on all raw agricultural commodities under FFDCA.
November 26, 1997	EPA announces PGS' request for a temporary "split" exemption of a tolerance for residues for Cry9C in corn for animal feed only under FFDCA.
April 10, 1998	EPA issues a final rule establishing the temporary "split" tolerance exemption for Cry9C.
May 22, 1998	EPA issues a final rule granting a "split" tolerance exemption to Cry9C.
April 7, 1999	EPA announces that AgriEvo USA had filed a petition to exempt from tolerance residues of Cry9C in or on all raw agricultural commodities.
December 21, 1999	EPA asks for input and comments on how to assess potential allergenicity of Cry9C in establishing "a reasonable certainity of no harm" before considering AgrEvo's request for a full exemption from tolerance under FFDCA.
February 29, 2000	EPA's Scientific Advisory Panel (SAP) meets to consider and define the issues and methods in determining food allergenicity of Cry9C.
April 5, 2000	A National Academy of Sciences (NAS) report states that Cry9C raises concerns of allergenicity
August 9, 2000	EPA announces new review process for plant-pesticide registraqtions, including Cry9C, in view of NAS recommendations.
October 31, 2000	EPA announces Anventis' submission of additional data and outlines review process.
December 5, 2000	EPA's SAP found

GOVERNMENT REVIEW AND APPROVAL PROCESS FOR STARLINK

StarLink's journey through the regulatory review and approval process was typical for a GM crop in the United States.[1] As with all crops bioengineered to

[1] The Coordinated Framework for Regulation of Biotechnology of 1986 establishes agency; responsibilities and regulatory policies for biotechnology products derived from existing statute, [Federal Register, June 26, 1986 (51 FR 23302)]. An in-depth discussion of regulatory issues in

contain an insecticide, StarLink required regulation under two different authorities: (1) the Federal Insecticide, Fungicide, and Rodenticide Act (FIFRA),[2] which requires EPA to register and label pesticides; and (2) the Federal Food, Drug and Cosmetic Act (FFDCA)[3], which requires EPA to establish a safe tolerance level for pesticides used on (or genetically engineered into) foods, and vests enforcement authority in the Food and Drug Administration (FDA).

Although health safety tests had found that Cry9C did not resemble any known allergens, results from other tests did not allow experts to completely rule out the potential for allergenicity. There were two particular concerns: one test showed that Cry9C protein could survive cooking or processing, and another determined that Cry9C is hard to digest. When in 1997 the original patent holder asked EPA to allow Cry9C corn to be registered for use only for livestock feed and industrial purposes, EPA exempted the plant-pesticide from the requirement for a tolerance in animal feed.thus granting it a so-called split registration. EPA required the company to take all actions needed to prevent StarLink from getting into the human food chain. Giving split registration is a common practice with conventional chemical pesticides, as each registration (or tolerance) specifies the crops on which use is allowed.

In 1999, the AgrEvo USA seed company, the next owner of the StarLink patent, asked EPA to set a tolerance for use in food products. EPA responded by asking for input from industry and academia on methods to ascertain Cry9C's potential for allergenicity and by convening a Scientific Advisory Panel in early 2000 to review the issue. Meanwhile, in April 2000 the National Academy of Sciences issued a separate report reiterating concerns about Cry9C's possible allergenicity and advising the EPA to improve the testing on the human and environmental impacts of Bt crops.[4]

---

food biotechnology can be found in CRS Report RL30198, Food Biotechnology in the Unites States: Science, Regulation, and Issues, by Donna Vogt & Mickey Parish.

[2] FIFRA (7 USC 136 et seq.) regulates the distribution, sale and use of pesticides in the United States. EPA defines registration as the formal listing of a pesticide before it can be sold and distributed in intrastate or interstate commerce. To register a pesticide the burden of proof is on the producer to demonstrate no "unreasonable" adverse effects on the environment.

[3] 3 FFDCA requires that the EPA establish a tolerance level for pesticide residues in raw agricultural commodities (21 USC 346a). EPA may determine that a pesticide (e.g., Cry9C \ is safe and exempt from the requirement of a tolerance if it finds, with "reasonable certainty"that aggregate exposure to residues will not cause harm.

[4] National Research Council. April 2000. Genetically modified pest-protected plants: Science and Regulation. National Academy Press. Washington, D.C. 260p.

HOW EVENTS UNFOLDED

On September 18, 2000, the Washington Post reported that tests ordered by a coalition of groups opposed to biotechnology had found traces of genetic material from StarLink in Kraft's taco shells in grocery stores in Washington, D.C. Kraft voluntarily recalled all taco shells from grocery stores 4 days later, after confirming the finding. Several other recalls by retailers have taken place subsequently, and in November the FDA exercised its enforcement authority by recalling over 300 corn products.

On September 26,2000, Aventis (the third and current owner of the StarLink patent) instructed its seed distributors in the United States to stop sales of StarLink seed corn for planting in 2001; shortly thereafter the company voluntarily agreed to cancel its EPA registration of StarLirik for feed and industrial use, thus taking the product off the market.

On October 25, Aventis submitted new safety information and asked EPA to grant a time-limited approval of up to 4 years for the presence of the corn in human food. According to Aventis, four years is how long it could take for all StarLink-contaminated corn to clear food channels.

On September 29, Aventis agreed with the U.S. Department of Agriculture (USDA) and EPA to purchase all of the 2000-crop year StarLink corn, offering farmers a 25-cent premium over the price of corn on October 2, which was $1.9925/bushel. Under the program, USDA will purchase the corn from the farmers and assure that it is distributed into feed and industrial channels only. Aventis will reimburse USDA for the cost. Early estimates of the cost to the company of implementing the buy-back program have ranged between $60 to $100 million and involve close to 80 million bushels.

International repercussions concerning StarLink began on October 24, 2000, when the Consumers Union of Japan found traces of the variety in snack foods and in animal feed. Under Japanese regulations, StarLink is not approved for any use and there is a zero tolerance threshold for StarLink in corn imports. Korea, the second largest market for U.S. corn, also has recalled corn products after finding StarLink traces in imported taco shells in early November. No immediate repercussions are expected in Europe, where imports of U.S. corn are small (0.1% of U.S. exports.)

THE CURRENT SITUATION

StarLink down on the farm. According to Aventis, farmers planted StarLink on 248,000 acres in 26 states in 1999. In 2000, farmers planted the variety on 352,000 acres in 29 states. These figures represent 0.32% and 0.44% of corn acreage in the United States in 1999 and 2000, respectively. Production estimates for StarLink are 38 million bushels (1999) and 54 million bushels (2000), which represent between 0.4% to 0.5% of U.S. crop production. However, in accordance with EPA rules, an additional 30 to 40 million bushels of non-StarLink corn harvested from buffer acres must be included in total production figures.[5] As of late October, Aventis had located all but 1.2 million bushels of the 2000 StarLink crop. The 1999 crop is so far along in the marketing chain that Aventis cannot determine its whereabouts. At the farm level, most producers already have sold their 1999 StarLink crop; they will be paid for the 2000 crop at the level set in the USDA- Aventis buy-back program (see above). StarLink in the market. Aventis maintains that about 12% of the 2000 crop, 9.6 million bushels, could be illegally present in food products. Aventis is working to forestall farther contamination by designating specific elevators for delivery and distribution into approved feed and industrial use outlets. However, the extent of commingling of StarLink with non-StarLink corn in food channels is harder to pin down. Media reports vary and high estimates abound because it does not take very much StarLink in a sample to result in a positive finding (see box on Testing). For example, experts at Iowa State University claim that a large proportion of corn in Iowa elevators may contain traces of StarLink.[6] Most experts agree that factors such as the handling of the corn from buffer strips and the possibility of StarLink pollen drifting into neighboring fields could elevate contamination estimates substantially. In addition, there is suspicion that the prevalent StarLink contamination in food currently on the shelves could be from the 1999 StarLink crop.[7] According to USDA, 1.8 billion bushels from the 1999 season remain in elevators as carryover. This represents close to 20% of that year's crop. StarLink exports. In early November 2000, USDA and Japan's Health Ministry implemented a plan to assure that no commingled corn is shipped to Japan. The plan provides that USDA will

[5] The 2000 StarLink Bt Grower Agreement states that: "In accepting StarLink corn, grower agrees to direct the harvested grain and grain grown within 660 feet of the StarLink grain towards domestic feed (e.g. animal feed) and/or non-food industrial purposes. Grower agrees not to use this grain for food use or allow it to enter grain export channels."

[6] Personal Communication with Dr. Charles Hurburgh, Professor at the Department of Agricultural & Biosystems Engineering. Iowa State University. November 2, 2000.

[7] Associated Press. November 3, 2000. Corn recall expands to stores, restaurants across country. Washington Post p.A5.

test for the presence of StarLink at domestic shipping locations, again on barges and railcars, and finally at export points. The protocol pertains to corn imports for food and feed use. Meanwhile, in response to a request from USDA and the U.S. grain industry, the Japanese Agriculture Ministry has agreed to review the safety of StarLink corn as an animal feed. The United States sold 600 million bushels of corn to Japan in 1999 with an estimated value of $1.45 billion. Sales to Japan represent 30% of U.S. corn exports. Late in 2000, USDA reports showed that corn sales to Japan were decreasing, but an agreement between the government of Japan and USDA in December promised to reverse the trend. Under the agreement, Japanese inspectors monitor tests of corn feed shipments and certify them as StarLink-free.

EMERGING ISSUES

Liability

The most frequently asked question is who is responsible for StarLink's illegal appearance in food products. EPA officials have said that Aventis, as a condition for its license, had the responsibility to ensure that the corn did not get into the human food supply. Aventis has said that it required farmers who grew StarLink to sign agreements to use the corn only for animal feed or industrial use and to treat non-StarLink corn harvested from buffer strips as part of the StarLink crop. In addition, Aventis has claimed that seed bags carried a label detailing these requirements. Many agree that the issue of liability is likely to be under judicial review for a long time, as farmers, elevators, processors, and others seek redress from losses that have been estimated in the hundreds of millions of dollars.[8]

Split Approvals for GM Crops by EPA

Critics have questioned EPA's decision to grant StarLink approval for animal feed and not for human food. Some are asking EPA to withhold approval of biotechnology crops until they have clearance to be used in food. These critics point to the StarLink incident as an example of how the "split" approval policy

---

[8] K.T. Arasu, Reuters News Service. "Anger in Iowa over gene-altered corn controversy." November 2,2000.

has gone wrong and threatens the confidence of consumers and foreign markets.[9] Others have argued that split registrations for GM varieties should be permitted on a case- by-case basis - for example, for those containing edible vaccines or for industrial-use-only cultivars, and where segregation from food varieties can be better ensured.

Mandatory Pre-Market Approval and Review

This incident has raised serious questions about the adequacy of the current system, especially in the areas of monitoring and enforcement. Some are calling for requiring a fully validated testing procedure for identifying DNA in crops and finished goods as a precondition to approval by EPA. Others are calling for a mandatory review of each new GM variety before it reaches the market by panels composed of state and federal agencies, industry, exporters, academia and other interest groups. Critics of these proposals argue that it would become increasingly difficult to register or introduce biotechnology innovations under such a system, and that the increasing genetic complexity of second and third generation GM crops would make testing for individual traits an expensive proposition.

LEGISLATION

Several bills were introduced in the 106th Congress which addressed the issue of mandatory labeling of GM foods.[10] In response specifically to the StarLink events that began in September 2000, Senator Durbin introduced a bill late in the 106th Congress that would have amended the Federal Food, Drug and Cosmetic Act to require pre-market consultation and approval for GM foods (S. 3184). The bill proposed to codify the existing system of voluntary consultations with the Secretary of Health and Human Services, giving FDA the authority to regulate

---

[9] See Kraft Foods (September 22, 2000) Official Statement; Grocery Manufacturers of America. (September 22, 2000) Press Release; National Corn Growers Association. (September 29,2000) "Hot of the Cob: Grains Council, NCGA Respond to Corn Concerns in Japan."

[10] Bills include H.R. 3377 (Kucinich) and S. 2080 (Boxer). For overview see CRS Report RS20507, Labeling of Genetically Modified Food, by Donna Vogt & Brian Jackson.

genetically engineered food and to require public participation in the decisions. Bill provisions would have also authorized FDA to test products on the market to determine whether unapproved genetically engineered materials are present.

Chapter 4

LABELING OF GENETICALLY MODIFIED FOODS[‡]

Donna U. Vogt

INTRODUCTION

Questions about the safety of bioengineered foods or genetically modified foods (GM foods) have led a variety of observers (consumers, Members of Congress, food companies, and international environmental negotiators) to support the labeling of GM foods. These foods were developed using recombinant DNA and related techniques to alter the genetic makeup of living organisms that are destined for human consumption. These techniques allow scientists to identify and isolate one organism's genes for useful traits and insert them into another plant or animal. Although some of the modifications made by bioengineering could be accomplished through traditional breeding methods, these newer techniques are more precise and predictable, can use genes from other species, and developers can create seeds having the desired traits faster than traditional methods. Currently, GM food crops planted and marketed by U.S. farmers include beets, corn, canola, cotton, rice, tomatoes, potatoes, soybeans and others. Each of these food products has been modified with added herbicide-tolerance and/or insect-resistance traits. Because of their unique nature, GM foods cross traditional

[‡] Excerpted from CRS Report RS20507.

regulatory regimes and involve several agencies with food safety mandates, including the FDA, the Department of Agriculture (USDA), and the Environmental Protection Agency (EPA), although only the FDA is in charge of regulating food labels.

Labeling of GM foods is being debated in Congress and in public forums. Two bills in the 106th Congress would have required mandatory labeling for all foods containing at least 0.1% ingredients of genetically modified organisms (GMOs).[1] In late 1999, FDA held three public meetings to solicit views on the most appropriate way to inform the public about GM foods. The views expressed at the meetings about labeling GM foods were described as polarized between those who wanted labeling and those who did not.[2] Bills are expected to be introduced in the 107th Congress that would mandate labeling.

Bioengineered foods shipped overseas will soon need documentation identifying them as such. On January 29, 2000, delegates from 140 governments finalized the Cartagena Protocol, known as the Biosafety Protocol.[3] It requires exporters of bioengineered commodities used for food, feed, or for processing to label shipments with a declaration that the product "may contain" GMOs and that the products are not intended for release into the environment. The Protocol does not require exporters to segregate bioengineered products from traditional products. Some say this documentation is not "labeling" but a shipping documentation. A State Department spokesman stated that the Protocol was "a step toward depolarizing the debate about biotechnology."[4]

Supporting and opposing views on the wider issues concerning the value and acceptance of GM foods are discussed in chapter 1. This chapter briefly describes

---

[1] One bill spelled out a system for each stage of food production whereby all persons with custody over the food (including the seed company, farmer, and food manufacturer) would have to label foods they know contain GMOs. If the food does not contain GMOs, they would have to issue a guaranty that the food does not contain GMOs. If the custodian of the food gave a false guaranty, there would be civil penalties up to $100,000 for each violation.

[2] Statement by Joseph A. Levitt, Director, Center for Food Safety and Applied Nutrition, Food and Drug Administration, Department of Health and Human Services before the Senate Health, Education, Labor, and Pensions Committee, September 26, 2000.

[3] On June 4, 1993, President Clinton signed the Convention on Biological Diversity, an international agreement negotiated at the 1992 Earth Summit in Rio de Janeiro under the auspices of the United Nations. This diversity agreement, ratified so far by 174 countries **but not by the United States**, calls for protecting a variety of plants and animals found in the wild. A second meeting in Indonesia in November 1995, at the Conference of the Parties to the Convention on Biological Diversity (COP-2), countries agreed to fund negotiations for a "biosafety protocol."

[4] Pollack, Andrew. "130 Nations Agree on Safety Rules for Biotec Food." *New York Times*, January 30, 2000. p. A1 and A6. Also see CRS Report RL30594, *Biosafety Protocol for Genetically Modified Organisms: Overview*, by Alejandro E. Segarra and Susan R. Fletcher. July 5, 2000.

FDA's food labeling policies and its policy rationale regarding the labeling of GM foods. It also discusses gives the views of supporters and opponents of labeling.

CURRENT FDA LABELING POLICIES

FDA is responsible for establishing standards for the labeling of all foods except for meat, poultry, and some egg products. The Federal Food, Drug, and Cosmetic Act (FFDCA) gives FDA broad authority to regulate foods by prohibiting the entry into interstate commerce of adulterated or misbranded foods. It is the legal responsibility of food manufacturers to produce foods that are not adulterated, unsafe, filthy, or produced under unsanitary conditions. The Act also requires that "food additives" not be marketed unless they have FDA's approval or are considered generally-recognized-as-safe (GRAS).[5]

FFDCA prohibits (in §403 Misbranded Food) statements in labels or labeling that are "false or misleading in any particular," and requires disclosure of "material" information on the label which gives the statement of identity or common/usual name of the food; the quantity of contents; the name and place of business of manufacturer, packer, or distributor; the ingredient information including any additives; and nutrition information. FDA does not require, however, prior approval for food labels nor does the FFDCA authorize FDA to require label warnings on food products, although the presence of a potential allergen must be included. FDA also gives informal advice to food manufacturers on labeling, if asked, and issues threats of enforcement actions against products whose labels fail to comply with regulatory requirements.

CURRENT FDA LABELING POLICY FOR GM FOODS

FDA determined in May 1992 that the scientific and regulatory issues posed by bioengineered food are not substantively different from those raised by non-bioengineered foods. Thus, FDA generally regulates GM foods no differently than foods created by conventional means. Nevertheless, FDA determined that there could be circumstances with GM foods that would require special review.[6] These

---

[5] Affirmation petitions for "generally-recognized-as-safe" (GRAS) status must be filed to gain FDA's formal agreement with a sponsor's independent determination that a substance is GRAS. A manufacturer may file a food additive petition even if FDA considers a substance GRAS.

[6] U.S. Food and Drug Administration. Statement of Policy: Foods Derived From New Plant Varieties. *Federal Register*, v. 57, no. 104, May 29, 1992. p. 22984-23005.

are:

- the gene transfer produces unexpected genetic effects;
- the levels of toxicants in the food are significantly higher than present in other edible varieties of the same species that have not been modified;
- nutrients in the bioengineered food differ from those in traditional varieties;
- the sources of the newly introduced genetic material come from a food plant associated with allergies found in humans;
- the food from the new variety differs significantly in composition from food of non-modified varieties;
- the food contains marker genes that theoretically could reduce the therapeutic effects of clinically useful antibiotics;
- the modified plants are developed to make substances like pharmaceuticals or polymers, as well as food; or
- the food is to be used for animal feed and has changes in nutrients or toxicants from the non-modified version.

An example of this special review process focused on the Flavr Savr tomato. In a effort to keep tomatoes fresher and longer on the vine, the Calgene Corporation genetically modified a strain of tomato to reduce activity of a particular enzyme (polygalacturonase) that affects softening of outer tissue during ripening. Because Calgene used an antibiotic marker as part of the modification process, a small amount of a non-tomato protein was produced in the Flavr Savr tomato. FDA viewed that protein as a food additive since it would not be found in an unmodified tomato and therefore, changed the tomato's composition. As a result, Calgene petitioned FDA and gained food additive approval for the protein in May 1994. Once approved, all labels listed the protein as an ingredient. In other cases, the composition of the GM product was deemed different enough from its unmodified counterpart to necessitate a different label. For example, FDA required the renaming of a soybean oil whose fatty acid composition had been altered by engineering. The new name, "high oleic acid " describes differences in the oil but not its production method.

On January 18, 2001, FDA published in the *Federal Register* a "Draft Guidance for Industry: Voluntary Labeling Indicating Whether Foods Have or Have Not Been Developed Using Bioëngineering" and is seeking comments. In this document, FDA reaffirmed that it would not require special labeling of all bioengineered foods because the use of bioengineering, or its absence, does not itself cause a material difference in the food. However, the agency did suggest that because of the strongly divergent views on labeling, manufacturers may consider providing more information about bioengineered food. The information given, however, must be shown to be truthful and not misleading. To avoid false or misleading statements about the absence of bioengineered ingredients (because there are no established threshold levels of bioengineered constituents or ingredients in foods), or to avoid implying that one food is superior to others, FDA suggests not using statements such as "GM free" or "biotech free." The agency does suggest the word "biotechnology" is preferred by some consumers over "genetic engineering" or "genetic modification." It also claims that if validated testing is available, it can used to verify whether the label is true. Or manufacturers could keep records to document the reasons why a food's label is truthful. On the same day, FDA published a proposed rule that would make mandatory the current voluntary consultation process whereby the industry would have to formally notify the agency 120 days before releasing a food product onto the market and would have to supply safety test data to FDA.

VIEWS ON LABELING OF GENETICALLY MODIFIED FOODS

Most consumers are very sensitive to the safety of the food they and their families eat and want to be confident that the food products they consume are wholesome and will cause them no harm. Most U.S. consumers trust FDA to ensure the safety of all foods whatever method was used to produce them. However, some concerned consumers, supported worldwide by various environmental groups including Greenpeace, question whether GM foods have been tested properly to ensure their safety from risks such as unknown allergens or toxins or for potential impact on the environment and biodiversity. They believe that government agencies should carefully review and approve the data from safety tests conducted by developers of GM foods. They also believe that foods produced through this technology could radically change the food supply of this country. These consumers and others have demanded the "right-to-know" which foods have been bioengineered. They want access to information on a label that would allow them to identify these products.[7] Supporters of labeling these foods claim that such labeling would allow them to have the knowledge to choose to avoid their purchase or consumption.[8] Other supporters compare this labeling with knowing how much fat is in a food product. To them, a label is not automatically a danger or hazard claim. They just want to know that the food they are eating was made through bioengineering.

Other consumers support mandatory labeling due to religious, ethical, or other strong personal values. These consumers, who routinely purchase specialty foods produced and labeled to abide by religious law or environmental concern, are accustomed to foods which identify their production process, i.e., certified kosher, halal, or organic production, on their label. These consumers want the GM production process identified so they can distinguish these products from those produced without the use of this technology. Labeling would allow these consumers the opportunity to choose to avoid certain foods.

Most seed producers, farmers, and food manufacturers oppose government mandated labeling requirements. They maintain that since companies producing seeds or plants already test products extensively for safety, and strive to maintain the reputation of their brands, labeling is unnecessary. They also claim that their compliance with FDA's 1992 GM labeling policy ensures that should the GM food contain any modification in nutrients, toxins, or allergens from traditional foods, it would be labeled as such.

Many who oppose mandatory labeling also claim that such a policy would constitute a significant restraint on trade if producers want to make a "GM-free" claim. As suggested in the draft guidance described above, FDA does not recommend this type of claim. To verify such a claim, commodities would have to be separated into two lots: those produced from GM seed and those produced from conventional seed. To provide accurate information and compliance,

[7] In 1996, in a ruling which applies only to three states, Vermont, Connecticut, and New York, the U.S. Court of Appeals for the Second Circuit held that food labeling cannot be compelled just because consumers want the "right-to-know" information. In overturning a Vermont law that required labeling of dairy products from cows treated with recombinant bovine somatotropin (BST), the court found that such regulation to satisfy the public's "right to know" is a constitutional violation of commercial free speech. The court wrote, "Were the consumer interest alone sufficient, there is no end to the information that states could require manufacturers to disclose about their production methods .... Absent, however, some indication that this information bears on a ... human health or safety concern, the manufacturers cannot be compelled to disclose it. Instead, those consumers should exercise the power of their purses by buying products from manufacturers who voluntarily reveal it." *International Dairy Foods Association, et al. v. Amestoy*, 92 F. 3d 67 (2nd Cir. 1996) and in Henry I. Miller, "Genetic Engineering," *Science*, v. 284, no. 5419, May 28, 1999. p. 1471-1472.

[8] U.S. Department of Agriculture's final rule for the labeling of organic foods will provide consumers with a food supply that does not contain GM foods.

elaborate and costly record keeping systems of segregating bioengineered foods would be needed at all stages of production (including seed development, planting, harvesting, distribution, processing, and packaging). Because of the pervasive use of bioengineered foods in food production, such segregation and traceability would also require extensive DNA testing at each step of the food production process and third party verification (either by the government or the private sector) of the final product. By imposing segregation and testing on the entire agricultural system, costs of providing this information would increase. In addition, the increase in the unit costs within the food system, might make small firms less viable and encourage industry consolidation.[9]

In September 2000, StarLink corn containing a Cry9C protein, a protein approved only for use in animal feed, turned up in taco shells. Without approval for human use, or exemption from approval, the Cry9C protein is considered an adulterant. The current regulatory system would still not necessarily have caught this adulterant in human food, even if there had been a mandatory labeling program. In December 2000, FDA published a sampling and testing guidance for the industry so that testing results could be used to verify the labeling of corn with or without the Cry9C protein. Other opponents of GM food labeling are concerned about the risk of "crowding out" or confusing other important information on the product label. Because the presence of too much information on a label makes it difficult to effectively transmit knowledge about potential concerns, they believe that adding what they deem as extraneous information on the label could have negative effects. Opponents of labeling also have hypothesized that consumers who know little about the issue might assume that any information included on a label constituted an implied hazard warning. As a result, if GM products are so labeled, these consumers might avoid them based on a judgement of implied risk.

Many industry groups support the current voluntary labeling policy towards GM foods which allows manufacturers to label voluntarily features that appeal to consumers, as long as they are truthful and not misleading, and could satisfy consumer desires for label information. However, these groups also claim that a label is limited as to the level of detail it can contain. Since different consumers care about different things and all information cannot be given on a label, some groups have *suggested* that information could be provided through a variety of sources. Some food industry groups are particularly interested in having FDA create guidelines and quantitative thresholds for "GM free" claims so this label

---

[9] "Biotech Crops Hold Potential to Increase Food Prices, says Fed." *SCI Policy Report*, January 3, 2000.

can be verified, can be considered truthful, and not somehow imply that GM foods may be unsafe. With such controversy surrounding consumer acceptance of these GM foods and their labeling, some industry groups are questioning whether they want to use GM products in their retail foods.[10]

[10] Food industry manufacturers always consider whether any new food has an obvious consumer benefit, makes the food function better, or would make a difference in sales.

Chapter 5

ADOPTION OF GENETICALLY MODIFIED AGRICULTURAL PRODUCTS*

Geoffrey S. Becker

SUMMARY

Since the first genetically engineered (GE) crops became commercially available in the mid-1990s, U.S. farmers have been rapidly adopting them in efforts to lower production costs and raise crop yields. A so-called second generation of GE commodities could shift the focus of agricultural biotechnology from the input side (i.e., farm production benefits) to the output side (i.e., consumer benefits). Even though only 12 different plants with GE traits have been approved for commercial use, at least 60% of all U.S. foods likely contain at least some GE material. That's largely because two of those plants — corn and soybeans, where farmers have widely adopted GE varieties — are used in many different processed foods. Because the policy debate over impacts of agricultural biotechnology can be controversial, the 108th Congress can be expected to continue close oversight of the issue.

* Excerpted from CRS Report RS21381. Updated June 2, 2003.

INTRODUCTION[1]

Since the dawn of agriculture, farmers and researchers have been modifying plants and animals to improve growth rates and yields, create varieties and breeds resistant to pests and diseases, and infuse special nutritional or handling characteristics. Such modifications have been achieved by finding and using only those plants and animals with desirable traits, by crossbreeding them with other organisms, through hybridization, and by other techniques. Almost all of today's commercial crops were developed through crossbreeding and/or hybridization.

Now, using recombinant DNA techniques, scientists can genetically modify plants and animals by selecting individual genes that carry a desirable trait (e.g., resistance to a pest or disease) from one organism, and inserting them into another that can be raised for food, fiber, pharmaceutical, or industrial uses.

Since the first genetically engineered (GE, sometimes called genetically modified or GM) crop varieties became commercially available in the mid-1990s, U.S. fanners have been rapidly adopting them in efforts to lower their production costs and raise crop yields. A so-called second generation of GE commodities could shift the focus of agricultural biotechnology from the input side (i.e., farm production benefits) to the output side (i.e., consumer benefits). Researchers are seeking to develop products offering enhanced nutritional qualities. GE farm products are being tested that carry new pharmaceutical and industrial-use traits. Such products will be livestock as well as crop-based.

CURRENT GE APPLICATIONS

In 2002, GE crops were planted on an estimated 145 million acres worldwide. Although the total number of countries growing such crops had reached 16 by 2002, most of the total acreage is highly concentrated among four crops (soybeans, corn, cotton, and canola) and four countries. About two thirds of global acreage was in the United States; nearly a fourth was in Argentina; and most of the rest was in Canada and China.[2]

---

[1] Sources for the introduction include various reports by the U.S. Department of Agriculture's (USDA's) Economic Research Service (ERS); the Pew Initiative on Food and Biotechnology; and the Biotechnology Industry Organization (BIO).

[2] International Service for the Acquisition of Agri-biotech Applications (ISAAA), *Global Status of Commercialized Transgenic Crops: 2002.* ISAAA is a private organization that promotes agricultural biotechnology. Although no GM crops are approved for commercial use in Brazil, an estimated 20-30% of its 40-42 million acres of soybeans were believed to be GM varieties in 2002, according to various USDA and private analysts.

In the United States, as of early 2003, 12 different plants with GE traits were approved for commercial use. Approximately 70% of all U.S. upland cotton, 80% of all soybean, and 38% of all U.S. corn acres are now being planted with GE seed varieties, according to USDA's National Agricultural Statistics Service (see **Table** 1). Weed and insect control are by far the most widespread applications of GE crops.

Table 1. U.S. Adoption of Major GE Crops, 1996 and 2003 (percent of total U.S. acreage)

	Soybeans	Upland Cotton	Corn
1996	7	17	4
2003	80	70	38

Source: USDA, various publications, including NASS. *Crop Production (supplement): Acreage,* various issues. Percentages are rounded.

Herbicide tolerant (HT) crops are engineered to tolerate herbicides that would otherwise kill them along with the targeted weeds. These include so-called "Roundup Ready" or HT soybeans, HT upland cotton, and to a lesser extent, HT corn.

Insect-resistant crops effectively have the pesticide inserted into the plants themselves to control insect pests for the life of the crop. Currently Bt *(Bacillus thuringiensis,* a soil bacterium) crops are the only ones in commercial use. They are most prevalent in upland cotton to control tobacco budworm, bollworm, and pink bollworm, and in corn to control earworm and several types of corn borers.

Others approved for commercialization include varieties of flax, papaya, potatoes, radicchio, rapeseed, rice, squash, sugar beets, tobacco, and tomatoes. However, these are either not on the market or planted very little. In addition, several non-crop products are available, notably in dairy production. Chymosin, a biotechnology-produced enzyme, is used widely in cheese production. Bovine somatotropin (BST; also known as bovine growth hormone or BGH) is a naturally occurring protein that can be produced in greater quantities through genetic engineering. The GE version of BST was first approved by the U.S. Food and Drug Administration (FDA) in 1993. More than 30% of all U.S. dairy cows are administered BST to boost milk production (by an estimated 10-15%).

U.S. FOOD PRODUCTS CONTAINING GE CROPS[3]

Even though only 12 different plants with GE traits have been approved for commercial use, at least 60% of all U.S. foods likely contain some GE material. That's largely because two of those plants — corn and soybeans, where farmers have widely adopted GE varieties—are used in many different processed foods. Federal biotechnology rules do not require separation and labeling of GE crops and foods, so long as they are substantially equivalent to those produced by more conventional methods (see page 5).

Soy-based ingredients include oil, flour, lecithin and protein extracts; corn-based ingredients include corn meal and corn syrups, used in many processed products. Canola oil (mostly imported from Canada, where GE-canola is grown) and cottonseed oil are used in cooking oils, salad dressings, snack foods, and other supermarket items. No GE produced animals are yet approved for human consumption, although cheeses may contain chymosin, and dairy products may have been produced from milk containing GE-BST.

Because most other government-approved GE crops are not being grown commercially, few other GE-derived foods are reaching consumers. Possible exceptions are some yellow squash and zucchini varieties, which few farmers are growing, and Hawaiian papayas, although most U.S. papayas are imported (non-GE) types from Brazil, Mexico, and the Caribbean. Calgene's FlavrSavr tomato, first marketed to consumers from 1995 to 1997, was withdrawn after Calgene determined that the varieties being grown were not of consistently high quality. Other GE tomatoes are not yet on the market GE potato varieties may have peaked several years ago at 2-3% of the market, but were discontinued by the seed developer in 2001 mainly after several fast food and snack food companies declined to buy them. Sugar beets, rice, flax, and radicchio all have not been commercially marketed despite government approval, presumably in part due to perceived producer or consumer uneasiness.

Analysts say some farmers are wary of planting GE crop varieties because they are worried that their customers do not want them out of perceived food safely and environmental concerns. Biotechnology supporters counter that such concerns are unfounded (scientific evidence has found that GE crop varieties are safe) and that foreign governments are simply using such concerns to maintain barriers to imports. These divergent viewpoints have been on display as

[3] Sources: Cornell University, Genetically Engineered Organisms Public Issues Education Project (GEO-PIE). The website can be accessed at: [http://www.comm.cornell.edu/gmo/who.html];

biotechnology companies (led by Monsanto) move toward commercialization of GE varieties of wheat in North America. Some producers want to plant the wheat when it becomes available; some others fear rejection by foreign customers of not only GE wheat, but all U.S. and Canadian wheat, out of concerns that even non-GE shipments might unintentionally contain some GE grain. The latter group wants developers and government regulators to wait for more market acceptance before releasing GE wheat varieties. Concerns about international market rejection also maybe why Canadian flax growers and U.S. sugar beet growers also are not commercially planting these crops, even though they are approved for unregulated release, according to experts.[4]

NEW GE APPLICATIONS[5]

As noted, the majority of GE varieties entering commercial use — even recent ones — have been for on-farm row crop applications, particularly weed and insect control. Such input-related uses are under development for other crops including wheat and alfalfa; apples, bananas, lettuce, strawberries, and eventually other fruits and vegetables; sugarcane; peanuts; sunflowers; and forestry products. Drought and frost tolerance, enhanced photosynthesis, more efficient use of nitrogen, and other agronomic traits are being introduced through genetic modification. Tomatoes that can be grown in salty soils otherwise unsuitable for agriculture, and recreational turf grasses that are herbicide tolerant, pest resistant, and/or more heat and drought-tolerant also are under development. On the horizon is a generation of GE products that developers claim will offer a range of "output" benefits, including, for example:

Human food qualities (nutraceuticals): oilseeds low in saturated fats; canola with high beta-carotene (antioxidant content); tomatoes with anti-cancer agents; grains with optimal levels of ammo acids; rice with elevated iron levels, and with beta-carotene, a precursor of Vitamin A ("golden" rice); other vitamin-enhanced crops; "low-calorie" sugar beets; increased sugar in strawberries and corn to improve flavor.

---

and Colorado State University's *Transgenic Crops: An Introduction and Resource Guide* at |http://www.colostete.edu/programs/lifesciences/TransgemcCrops/iiidex.html].

4 GEO-PIE; also see *GM Wheat, A Sensitive Subject,* November 26, 2002, at Pro Farmer website: [http://www.agweb.com/news_show_news_article.asp?articleID=93223&newscat=GN]; "Farmers not sure they want genetically modified wheat," in the May 4, 2003 *Minneapolis Star Tribune;* and *Biotechnology and Agricultural Trade* in the CRS Trade Electronic Briefing Book.

5 Sources: ERS, *Economic Issues in Agricultural Biotechnology* (AIB-762), February 2001 (table, page 19); BIO; and Colorado State University.

Processing qualities: colored cotton; improved cotton fiber; high-solids tomatoes and potatoes; delayed ripening melons, strawberries, raspberries, and other produce (such tomatoes already are approved); wheat with altered gluten levels to improve baking; naturally decaffeinated coffee.

Specialty chemicals (plants as bioreactors): pharmaceuticals, antibodies, and medical vaccines such as diarrhea vaccines from bananas; blood proteins from potatoes; rabies vaccine and monoclonal antibodies from corn.

Transgenic livestock: pharmaceuticals from cow's, pig's or sheep's milk, such as a blood anti-coagulant, a substance to treat cystic fibrosis, and a human milk protein to use as a nutritional supplement; livestock with more rapid growth, less fat, disease resistance, and longer useful lives; GE salmon, (and later tilapia, trout, and flounder) that require as little as half the time to grow to market size.

REGULATION

The basic federal guidance for regulating biotechnology products is the **Coordinated Framework for Regulation of Biotechnology** (51 *Fed. Reg.* 23302), published in 1986 by the White House Office of Science and Technology Policy (OSTP). A key principle is that genetically engineered products should continue to be regulated according to their characteristics and unique features — not according to their production method. The Framework provides a regulatory approach intended to ensure the safety of biotechnology research and products, using existing statutory authority and previous agency experience with traditional breeding techniques.

The three lead agencies are USDA's **Animal and Plant Health Inspection Service (APHIS),** which regulates the importation, interstate movement, and field testing of GE plants and organisms that are or might be plant pests under the Plant Protection Act (PPA; 7 U.S.C. §7701 *et seq.)* and which regulates animal biologies (i.e., viruses, serums, toxins for animal vaccines) under the Virus, Serum, and Toxins Act (21 U.S .C. 151 *et seq.)*; the **Food and Drug Administration (FDA),** which regulates food, animal feed additives, and human and animal drugs, including those from biotechnology, primarily to ensure that they pose no human health risks, mainly under the Federal Food, Drug and Cosmetic Act (FFDCA; 21 U.S.C. §301 *et seq.)* and the Public Health Service Act (42 U.S.C. §201 *et seq.)*; and the **Environmental Protection Agency** (EPA), which must approve the use of all pesticides, including those genetically engineered into plants.[6]

The U.S. regulatory approach contrasts with that of many major trading partners: the European Union (EU), Japan, South Korea, China, Australia, and New Zealand either have or are establishing separate mandatory labeling requirements for products containing genetically modified ingredients, for example. These differing regulatory approaches (where consumer and some official attitudes toward GE foods are more skeptical) have begun to disrupt U.S. exports and heighten trade tensions (see below).

ISSUES FOR CONGRESS

U.S.-EU Dispute

On May 13, 2003, the United States and several other biotechnology-adopting countries announced they were filing a case before the World Trade Organization (WTO) against the EU's de facto 5-year moratorium on approving new products of agricultural biotechnology. The Administration had been under increasing pressure in recent months from key Members of Congress and farm groups to launch such a case. U.S. agricultural groups contend that not only have EU policies blocked their exports to the EU, but they have also fueled unwarranted concerns about the safety of agricultural biotechnology throughout the world, particularly in developing countries that, they assert, would benefit greatly from GE crops. EU officials say they have been moving as quickly as possible to reinstate biotechnology approvals while trying to reassure their consumers regarding safety issues — in part through the adoption of new labeling and traceability regulations (which U.S. farm interests assert will also prove unworkable and discriminatory against their products).[7]

Other Issues

Among other biotechnology issues are:

---

[6] For background see CRS Report *RL30198, Food Biotechnology in the United States: Science, Regulation, and Issues.*

[7] See: *European Union Moratorium on Biotech Crops* in the CRS Agriculture Policy Electronic Briefing Book.

- The impacts of GE crops on the environment (for example, can/do GE crops reduce soil erosion through more no-till practices, and decrease use of chemical pesticides), and on biodiversity (e.g., might GE varieties threaten other crops and other animals);
- Questions among some consumer groups about the safety of GE foods, and about consumer acceptance, including the issue of GE labeling;
- The extent to which a few large companies are gaining control over GE seed varieties — and how that might affect their cost and availability to fanners, and acceptance overseas;
- Support for and conduct of agricultural biotechnology research, including development of new crop and animal products and testing of their impacts (agricultural research in the past largely was publicly sponsored but much of the work in biotechnology now is privately funded);
- Issues surrounding intellectual property and patent ownership rights;
- Whether the current U.S. regulation and oversight of biotechnology remain appropriate, particularly as newer applications (e.g., biopharmaceuticals) emerge that were not in development when the regulatory regime was first established;
- The prospects for agricultural biotechnology to improve (according to proponents) or undermine (according to opponents) food security in developing countries and elsewhere in the world.

In the 107th Congress, lawmakers included, in the 2002 farm bill (P.L. 107-171), provisions aimed at supporting use of GE farm products. For example, Section 3204 creates a "Biotechnology and Agricultural Trade Program" authorizing appropriations of up to $6 million annually for technical assistance and grants to remove or mitigate significant foreign regulatory nontariff barriers to U.S. agricultural exports, including GE products. Section 10802 authorizes a science-based public education program on the use of biotechnology for producing human food. Separate legislation (P.L. 107-210) providing for trade promotion ("fast track") authority calls on U.S. officials negotiating trade agreements to press for the elimination of unjustified restrictions, including labeling, for biotechnology products. The 108th Congress is likely to continue to support biotechnology in U.S. agriculture, although controversy over impacts and benefits is expected to persist, suggesting ongoing oversight of the issue and, potentially, proposed bills to more stringently regulate its use.

Chapter 6

THE "TERMINATOR GENE" AND OTHER GENETIC USE RESTRICTION TECHNOLOGIES (GURTS) IN CROPS*

Alejandro E. Segarra and Jean M. Rawson

ABSTRACT

This chapter provides general background on "Terminator ge**Error! Bookmark not defined.**nes" in seeds and briefly examines three issues associated with these new types of biotechnologies: whether they influence vertical integration in the seed industry, whether research and development of "Terminator-type" genes should be funded by the U.S. government under Cooperative Research and Development Agreements (CRADAs), and whether food security and biodiversity of developing countries could be impacted. Issues of CRADA funding relating to domestic or international competitiveness are addressed by H.R. 209, as engrossed, and S. 804 in the 106th Congress.

Plant Genetic Use Restriction Technologies (GURTs) are a group of complex genetic transformations that insert a genetic "on-off switch" in plants to prevent the unauthorized use of genetic traits contained within. GURTs are one of the

* Excerpted from CRS Report RL30278.

latest bioengineering products developed for agriculture, with over 30 patents issued in the United States and Europe since 1997. The current state of biotechnology may allow GURTs to be introduced into the commercial seed market in the next 5 to 7 years. Controversy over GURTs first arose in 1998 when the U.S. Department of Agriculture and Delta & Pine Land Company received a U.S. patent for the rights to use a new genetic process which blocks the production of fertile seeds in genetically engineered plants. Detractors dubbed this utility patent "the Terminator gene" and mounted an extensive campaign against its use. In general, opponents view the use of GURTs as an accelerator of seed industry consolidation and vertical integration, and link the technology to potential adverse effects on food security and crop biodiversity worldwide. Others have questioned the role played by the federal government in funding and developing this kind of technology. Proponents in USDA and biotechnology firms argue that this new technology insures against illegal use of their intellectual property and helps to create new industry incentives. They further maintain that the level of investment protection achieved through GURTs gives growers in developing countries better and expanded seed choices, and the opportunity to move away from subsistence farming and into modem production agriculture. USDA has asserted that its role as an active participant in GURT development will ensure that the public interest is represented. An issue before Congress is whether current laws are adequate to assure that biotechnology innovations are applied in ways that do not have unwelcome economic, social or environmental side effects. This chapter examines the latest technical advances in Genetic Use Restriction Technologies and discusses key issues related to their potential impact on farming and agriculture. Plant Genetic Use Restriction Technologies (GURTs) are a group of complex genetic transformations that insert a genetic "on-offswitch" in plants to prevent the unauthorized use of genetic traits contained within, i.e., GURTs ensure that farmers cannot replant genetically modified crops by simply saving seed. One view is that the use of GURTs provides biotechnology companies with a reliable "turn-key" mechanism to ensure against the illegal use of intellectual property through a biological lock. However, another set of views looks at GURTs as potential threats to the environment and to farmers in developing countries. Since 1997, over 30 GURT patents have been issued in the United States and Europe; however, the technology is not expected to appear in the commercial seed market for at least 5 years.

In March 1998, the U.S. Department of Agriculture's Agricultural Research Service (ARS) and Delta and Pine Land Co. (D&PL), a major developer and seller of cotton seed, received U.S. Patent 5,723,765 entitled "Control of Plant Gene Expression." This patent gives the holders rights for the use of three new gene sequences that block the production of fertile seeds in genetically engineered plants. According to the patent holders, this so-called "Technology Protection System" (TPS) would ensure intellectual property protection for investments in genetic engineering and help create incentives to develop new plant varieties that satisfy changing market demands.

Shortly after the patent's issuance, the Rural Advancement Foundation International (RAFI), an organization headquartered in Winnipeg, Canada and Pittsboro, N.C., that is concerned about issues of crop genetic diversity and the impact of intellectual property rights on agriculture, began to organize opposition to the patent. RAFI coined the term "Terminator Gene" to label the technology as a negative one for farmers. Issues highlighted by RAFI were: the accelerating effect of these technologies on seed industry consolidation; the potential adverse effects of TPS on food security in the developing world; and the potential for worldwide losses in crop biodiversity. These issues have received wide press coverage, and other groups have joined RAFI in accusing ARS and the seed biotechnology industry of developing TPS with the intent of hastening monopoly control over global agricultural production.[1]

Concerns about the use of GURTs and seed sterility are now beginning to extend into the public policy arena. For example, on July 16, 1999 and acting on farm groups' concerns, the State of New Hampshire enacted a law to establish a 3-year study to examine "the threat to bio-diversity resulting from Terminator technology."[2] However, seed sterility is not the only biotechnology-based form of intellectual property protection under development and patents have been issued on other types of GURTs, as examined below.

In general, GURTs are a form of protection against the illegal use of a plant variety based on genetic engineering of the plant rather that on license agreements or end-user contracts. Companies, such as the Monsanto Corporation, typically license their biotechnology products under certain conditions. Some of these conditions include agreements by the farmers who purchase their seed to use them solely for planting a single commercial crop and agreeing not to save any seed produced for replanting or trade. Monsanto has repeatedly taken farmers to court for violations to these agreements.

---

[1] RAFI. Terminator seeds rejected by global network of agriculture experts. News Release-November 2,1998 (http://www.rafi.org/pr/release23.html).

[2] New Hampshire State General Court. HB 291. Adopted by both Houses on July 1, 1999 and signed by the Governor on July 16, 1999. Chapter 0282.

A recent report, produced for the Convention on Biological Diversity (CBD), breaks GURT patents into two major categories.[3] Those that deny unlicensed access to the entire plant variety or "Variety-level Genetic Use Restriction Technologies" (V-GURTs); and those in which only the "value-added" transgenic trait is protected from illicit use, or "Trait-specific Genetic Use Restriction Technologies" (T-GURTs).[4] Currently, V-GURTs account for most of the patents, but the line separating these categories is often blurred by the absence of current applications and many patents broadly describe GURTs which could be used in one way or another.[5] [6]

Both V-GURTs and T-GURTs feature metabolic "locks," which can be opened or closed only through the application of key "inducer" substances to the plant. Substances, such as plant-growth regulators, antibiotics, or herbicides are being researched as possible "genetic inducers" and will be under the proprietary control of the biotechnology companies. Most experts agree that it is unlikely that any GURT-containing crops will be available for at least 5 years. No GURTs have, to date, been produced or even left the scientist's bench for testing or trials. According to most specialists, the required technological breakthroughs needed to produce stable genetic expression and commercially viable GURT plants will require a very large undertaking. The process has to be tailored and tested individually for each variety and crop. In addition, the sheer complexity and unpredictability of present day genetic manipulation technologies greatly restrict the number of varieties that could be transformed.

In theory, any crop species could potentially be transformed to contain value-added genetic traits and GURTs. However, given the great number of technical barriers and the expected high cost of development, GURT-containing varieties will be limited initially to a small number of crops. These will likely include many of the most widely used crops, such as rice, wheat, cotton and soybeans. Another factor influencing the selection of crops for GURT transformation is the presence

---

[3] CBD: Subsidiary Body on Scientific, Technical and Technological Advice (SBSTTA). 4th Meeting. Montreal, June 21-25, 1999. Item 4-6 of the Provisional Agenda.

[4] Many such "value-added" genes are currently on the market, including genetic modifications for insect and virus resistance, herbicide tolerance, production of specialty oils, controlled ripening, low lignin, and others. Seed companies currently protect these traits by patents.

[5] One example of a V-URT is USDAD &PL 'TPS' patent. Other V-GURT patents had been issued previously, including: the Novartis Corp. (US-5650505 issued July 22, 1997), and the Monsanto Company (WO-9744465 issued Nov. 27, 1997).

[6] Examples T-GURT patents are Zeneca Ltd's (UK), entitled "Cysteine protease promoter from oil seed rape and a method for the containment of germplasm" (WO-9735983) issued on October 2, 1997; and Purdue Research Foundation's "Selective expression of Genes in Plants" (WO-9911807) issued on March II. 1999.

of current seed consumption trends that indicate continued declines in the use of saved-seed for replanting.

An issue before Congress is whether current laws are adequate to assure that the benefits of biotechnology innovations help rather than hinder the public trust. The concerns raised below represent salient issues and are, to a large extent, an important manifestation of the broader economic, social and environmental concerns surrounding the use and development of agricultural biotechnology.

SEED INDUSTRY CONCENTRATION AND VERTICAL INTEGRATION

While crop biotechnology is increasingly considered one of the most important agricultural innovations of this century, it has raised some concerns in the context of the current trends towards seed industry consolidation. Leland Swenson, president of the National Farmers Union, believes that recent developments such as biotechnology and the "Terminator gene" will hasten the process of vertical integration in agriculture and place the farmer at the mercy of agribusiness- conglomerates. Another concern is about "backward integration," a situation in which farmers might be forced to buy seed from one company in a conglomerate because the processing firm to which they wish to sell their crop requires specific genetic traits found only on those seeds. GURTs, according to this view, will greatly accelerate the already fast pace of vertical integration in seed companies.

That concern is not shared by those who view industry consolidation as part of a long continuing process in American agriculture. They contend that the economies of scale and the production efficiencies brought about by consolidation are naturally justified in view of the high costs of investment and of the high value of the results. Mark Drabenstott, director of the Center for the Study of Rural America, views consolidation in U.S. agriculture as a positive trend, and one that has led to lower-priced, higher quality food products for consumers and a leaner industry better able to compete in the global market place.[7]

Regardless, many U.S. farmers appear to have adopted the type of value-added, genetically modified (GM) crops that may eventually get GURTs. According to the USDA Economic Research Service (ERS), the adoption of

[7] Testimony before the U.S. House of Representatives, Committee of Agriculture hearing on Agribusiness Consolidation. February 15, 1999.

genetically engineered crops by U.S. farmers has increased rapidly.[8] For example, GM corn increased from 3.5 to 23.7 million acres between 1996 and 1998 (i.e., from 4.4% to 36% of total planted area). In other crops, such as soybean and cotton, increases have been even larger, with GM varieties accounting for 44.8% and 52.8% of total 1998 acreage, respectively. In addition, U.S. farmers nowadays depend less on saved seed for their planting. According to ERS figures, soybean grown from saved seed decreased from 23% to 14.9% from 1990 to 1997. Similarly, cotton planted from saved seed dropped from 26.5% to 17.8% between 1991 and 1997. In U.S. corn, the longstanding availability of high yielding hybrid seed (whose beneficial traits do not all reappear in a second generation) have virtually eliminated saved seed as a farming practice in corn production.[9] Presently, seeds for GM corn, soybean, canola and cotton are distributed through dozens of brand varieties, but the underlying biotechnologies are still restricted to a handful of trademarks owned by the Monsanto Company, Mycogen, AgrEvo GmbH, and the Novartis Corporation.

Large U.S. farm organizations have generally welcomed biotechnology as part of the agribusiness revolution in America Michael Yost, president of the American Soybean Association, considers biotechnology ". . . an unqualified triumph of modem science by, so far, reducing production costs and increasing the crop's value." According to Mr. Yost, soybean farmers understand the legitimate motivations of biotechnology companies to protect their investments by using license agreements or (eventually) GURTs. For him, the concern is that seed from generally less expensive conventional varieties might continue to be available, and that farmers are treated fairly under licensing agreements, end-user contracts, and other such devices.[10] Others, although supporting licensing agreements, contend that it is important to maintain a business environment that fosters competition and that facilitates the entry of new biotechnology firms with new products, thus providing farmers with a continuous broad array of seed choices.

[8] USDA-Economic Research Service. Genetically Engineered Crops For Pest Management. (http://www.econ.ag.gov/whatsnew/issues/biotech). July 1999.

[9] Dr. William McBride. USDA-ERS. Personal communication. July 6, 1999.

[10] Testimony before the U.S. House of Representatives, Committee on Agriculture. Subcommittee on Risk Management, Research, & Specialty Crops. Hearing on Agricultural Biotechnology. March 3, 1999. Personal telephone communication, July 7, 1999.

ROLE OF PUBLICLY FUNDED RESEARCH IN THE DEVELOPMENT OF GURTS

USDA's development of "Terminator-type" technology has fueled a controversy over federal funding of GURT research. Margaret Mellon, director of agriculture and biotechnology programs at the Union of Concerned Scientists (UCS), has charged that USDA's research agenda is insufficiently focused on the public interest. Mellon has also argued that the Department has a direct conflict of interest in the development of these types of biotechnologies while simultaneously being one of its principal regulators.[11] UCS believes that this type of state-sponsored research solely benefits the large biotechnology corporations.[12] RAFI has also accused USDA of not representing the public interest and has called on the Department to renounce its part of the TPS patent.

USDA is the joint holder of the TPS patent under the terms of a Cooperative Research and Development Agreement (CRADA) signed with Delta & Pine Land (D&PL). Congress authorized CRADAs in the Federal Technology Transfer Act of 1986 (P.L. 99-502) to enhance the ability of federal research laboratories to work with industry to commercialize technology. This act and various other federal laws—including the Stevenson-Wydler Act of 1980 (P.L. 96-480) and the Bayh-Dole Act of 1980 (P.L. 96-517)— make the transfer of new technology to the private sector and industry a responsibility of all federal research agencies. Currently, USDA and D&PL are negotiating license agreements for the use of the TPS patent. USDA-Agricultural Research Service (ARS) officials have declared that, as these negotiations evolve, ARS will be an active participant in deciding how the technology is applied, and that ARS involvement will ensure that the public interest is represented. According to these officials, it is ARS policy that technology in which it has an ownership interest will be made widely available, and it has announced that TPS will be available for research purposes to public and private researchers. In line with ARS policy, D&PL has agreed to make the technology widely available for sublicensing to other seed companies. The statutes governing the use of CRADAs guarantee to the industrial partner, such as D&PL, the option to choose, at a minimum, an exclusive license for a pre-negotiated field of use to the resulting invention made in whole or in part by a

[11] A thorough discussion of regulatory issues concerning food biotechnology in the U.S. can be found in CRS Report RL30198, Food Biotechnology in the United States: Science, Regulation, and Issues, by Donna Vogt & Mickey Parish.

[12] B.A. Baker, A new advisory panel will help USDA tackle the thorny issues raised by agricultural biotechnology. Bioscience. Vol 49: 438. June 1999.

government laboratory employee. CRADAs also protect the government's right to use the invention for its legitimate needs. ARS maintains that, in the spirit of the Federal Technology Transfer Act, its scientists have used CRADAs to expand expertise and speed development of many new technologies of direct benefit to farmers and consumers as well as to multinational biotechnology firms.

FOOD SECURITY IN DEVELOPING NATIONS

Advocates of GURTs and other use restrictions claim that these technologies will help stimulate private efforts to breed and market new and improved crop varieties in countries which have not benefited from these advances. For them, GURTs are only the latest in a long list of ways of compelling farmers to buy seeds from the companies that have developed them. They also maintain that this type of restriction against unlicensed use of their products is perfectly reasonable for most researched goods, including commonly available hybrid seeds. Ultimately, they affirm, GURTs and other restriction technologies will give growers in developing countries the option of moving from subsistence farming into modern production agriculture.

Supporters also argue that GURTs will ensure that U.S. farmers will compete on a more level playing field with respect to other farmers worldwide. North American farmers, they maintain, have been paying for seed biotechnologies, while at the same time, some of these same technologies have been pirated into other countries without payments by farmers receiving their advantages.

On the other hand, critics of GURT use argue that these technologies will result in unacceptably limited seed choices for farmers. Patrick Mooney of RAFI, maintains that marketplace competition will force farmers to choose only seeds that promise the highest crop yields, and which are often genetically modified (GM) varieties. Farmers could become dependent on the biotechnology companies for an ongoing supply of the GURT seeds. "What happens if a company decides one area is not a profitable marketplace?" asks Mooney. Moving away from crops with viable seeds could threaten developing countries' ability to insure a reliable food supply from one year to the next, he says.

In a recent speech to Monsanto executives, Gordon Conway, president of the Rockefeller Foundation, called upon the biotechnology industry to disavow the use of "Terminator technology.[13] Conway contends that current legal mechanisms

[13] Gordon Conway, President Rockefeller Foundation. Speech to Monsanto Board of Directors. Washington, D.C. June 29, 1999

and international trade agreements adequately protect against unlicensed use of patented plant varieties. For example, Conway asserts that the plant variety protection afforded by the International Union for the Protection of New Varieties of Plants (UPOV) is adequate against commercial pirating, and that UPOV qualifies as an acceptable crop intellectual property system under the World Trade Organization (WTO) Trade-Related Aspects of Intellectual Property Rights (TRIPS).[14]

Terminator-type technology that causes seed sterility, he declares, could cause widespread negative public reaction against biotechnology in general if farmers, unaware of the characteristics of purchased "Terminator gene" seed, attempt to re-use it. The issue of "dead seeds" and the potential loss in food security are seen as legitimate issues of national security by many governments.

Recently, in a move expected by some biotechnology analysts, Monsanto sided with Mr. Conway's views. In an open letter from Monsanto's CEO, Robert B. Shapiro, to the Rockefeller Foundation's president. Monsanto announced that it would not pursue or commercialize "Terminator gene" technology.[15] Nevertheless, Mr. Shapiro did not rule out the possibility of using other technological approaches to gene protection that do not render seeds sterile, such as trait level-GURTs, as a means to protect and gain a return on investments in biotechnology innovation.

Experts examining GURTs for the Convention on Biological Diversity suggest that the greatest risk to food security is associated with wide adoption of Variety level-GURTs (as opposed to Trait level-GURTs). V-GURTs, they argue, may lead to increased farmer dependence on seed from a few commercial suppliers, and lead to increased levels of vulnerability during war, social unrest or natural disasters. T- GURTs, they assert, could contribute positively towards responding to private sector concerns of secure investments, while notably increasing farmers' choice on the decision to activate the trait if they so choose.

Some in the industry, like the Monsanto Corporation, have agreed that concerns about GURTs should be heard and carefully considered before any decisions are made to commercialize them. One important issue which must be tackled, according to Monsanto, is that of small farmers who rely on saved seed to provide seed stock for next year.

[14] The TRIPS Agreement was concluded at the Marrakesh, Morocco WTO conference on April 15, 1994 and came into effect on January 1, 1995.

[15] Shapiro, R.B. Open letter from Monsanto Open Letter From Monsanto CEO Robert B. Shapiro To Rockefeller Foundation President Gordon Conway. October 4, 1999. http://www.monsanto.com/monsanto/gurt/default.htm

Crop Biodiversity

Another important set of issues relates to the impact of GURTs on genetic diversity in crops. One common concern is that GURTs could limit the sharing of germplasm among plant breeders worldwide. This potential for "locking up" of useful genetic variation, they argue, would be contrary to the spirit and intent of the Plant Variety Rights convention adopted in Paris in 1961. One reaction by the Consultative Group for International Agricultural Research (CGIAR), an important public source and repository of germplasm worldwide, has been to adopt a "non-use policy": CGIAR will not incorporate into its breeding materials any genetic systems designed to prevent seed germination.[16] For its part, USDA has announced that it has no intention to introduce TPS into any germplasm in their National Plant Germplasm System (NPGS) collection or into its crop development research. CGIAR, with its 16 affiliated International Agricultural Research Centers, and USDA-NPGS are considered among the largest public germplasm repositories in the world.

GURT critics also argue that farmers, who already face a small pool of genetically uniform varieties, will face an even more restricted genetic make-up in their crops due to GURTs' faculty for stopping natural gene exchange processes, such as cross pollination and the formation of locally adapted crop varieties (i.e., land races). They caution against the risk of allowing such limited gene pools, as crops may become more susceptible to new diseases or pests.

Some scientists and GURT advocates who consider these negative scenarios unlikely, respond that there is no correlation between GURTs and lack of genetic diversity. They argue that GURTs should increase incentives for many private seed companies to breed crops which have not received sufficient attention in the past, and that it is likely that genetic diversity will increase as breeders focus on providing unique and improved versions of germplasm to farmers.

Concern has also been raised about accidental transfers of GURT gene(s), particularly in open pollinated species such as corn, which could lower the productivity of nearby conventional (i.e., non-GM) fields. USDA, for its part, says that the TPS is intended for deployment only in self-pollinating crops to ensure that the risk of pollen transfers to conventional fields remains extremely low.[17] One legislative proposal that potentially affects CRADAs (such as the one that

---

[16] CGIAR has issued no policy position regarding the use of T-GURTs. Personal communication with Dr. Manuel Lantin, Science Advisor. CGIAR Secretariat. Washington D.C. July 2,1999.

[17] Knipling, E.B., -USDA-ARS Associate Administrator.- Letter to the United Nations Environmental Programme. May 14, 1999.

resulted in "Terminator-type" technology) has been under consideration in the 106th Congress. Section 6 of the proposed Technology Transfer Commercialization Act of 1999 (H.R. 209, as reported, and S. 804), would amend the Stevenson-Wydler Technology Innovation Act of 1980 to, among other things, require federal agencies with laboratories that are engaged in CRADAs to provide to Congress a review of their general policies and procedures for entering into major CRADAs involving critical national security technology, or that could impact domestic or international competitiveness.[18] Future USDA CRADAs — particularly those dealing with biotechnology such as TPS and GURTs— could potentially be required to undergo this review process.

[18] A detailed discussion on CRADAs in U.S. research policy is found in CRS Report 98-862, "R&D Partnerships and Intellectual Property', and Issue Brief 89056, "Cooperative R & D: Federal Efforts to Promote Industrial Competitiveness" by Wendy Schacht.

Chapter 7

AGRICULTURAL BIOTECHNOLOGY: THE U.S.-EU DISPUTE[†]

Geoffrey S. Becker and Charles E. Hanrahan

INTRODUCTION

The United States and several other countries in May 2003 initiated a formal challenge before the World Trade Organization (WTO) of the European Union's (EU's) de facto moratorium on approving new agricultural biotechnology products, in effect since 1998. The moratorium reportedly has cost U.S. corn growers some $300 million in exports to the EU annually. U.S. growers plant genetically engineered (GE) varieties mainly for weed and pest control. They do not segregate them from non-GE varieties, because the U.S. regulatory system recognizes them (once approved for commercialization) as substantially equivalent to non-GE varieties. The EU moratorium, U.S. officials contend, threatens not only other agricultural exports to the EU, but also exports to other parts of the world where the EU approach to regulating agricultural biotechnology is taking hold. The EU approach presumes that the products of biotechnology are deemed to be inherently different than their conventional counterparts and should be more closely regulated.

[†] Excerpted from CRS Report RS21556. Updated July 3, 2003.

OVERVIEW[1]

On May 13, 2003, the U.S. Trade Representative (USTR) and the U.S. Secretary of Agriculture announced that the United States, Canada, Argentina, and Egypt were filing a case before the World Trade Organization (WTO) challenging the European Union's (EU's) de facto 5-year moratorium on approving new agricultural biotechnology products. (Egypt subsequently decided against filing with the United States.) U.S. agricultural interests contend that these policies have not only blocked their exports to the EU, their fourth-largest foreign market, but also fueled unwarranted concerns about the safety of agricultural biotechnology throughout the world.

EU officials say they have been moving as quickly as possible to reinstate biotechnology approvals while trying to reassure their consumers regarding safety issues. There is concern among some policymakers that the filing of a WTO case will only escalate trade tensions between the United States and the EU, where the value of two-way agricultural trade amounts to more than $14 billion annually.

BACKGROUND

The United States accounted for two-thirds of the 145 million acres planted to GE crops in 2002, according to the International Service for the Acquisition of Agri-biotech Applications (ISAAA), a group supportive of the technology. In 2003, approximately 80% of all U.S. soybean, 70% of all U.S. cotton, and 38% of all U.S. corn acreage, were planted with genetically engineered (GE) seed varieties, designed mainly to control pests (weeds and insects). Once GE varieties are approved, the U.S. marketing and regulatory system generally does not distinguish between them and non-GE varieties and their products (except for products labeled "organic," where genetic engineering is prohibited).

Outside of the United States, however, gaining access to markets like the EU, where consumers and their governments have been more wary of biotechnology, has been difficult. With minor exceptions, the EU has approved no agricultural biotechnology products since 1998, even though it has an elaborate approval process in place, and, in October 2002, it implemented revisions to that process

[1] Sources include: USTR. *2003 National Trade Estimates Report on Foreign Trade Barriers;* Pew Initiative on Food and Biotechnology, *U.S. vs. EU: An Examination of the Trade Issues Surrounding Genetically Modified Food,* 2002; European Commission, *Questions and Answers on the regulation of GMOs in the EU,* March 4, 2003; and *Biotechnology Regulations and the WTO,* International Agricultural Trade Research Consortium, Working Paper #02-2, 1/02.

aimed at reassuring its member states and the public about the safety of its regulatory system. More than a dozen products in early 2003 were awaiting approval, some for as long as 6 years. A majority of EU states has effectively blocked the release of any new GE crops into the environment. These states—France, Belgium, Luxembourg, the Netherlands, Germany, Austria, Italy, Ireland, Greece, Spain, Portugal, and Finland — say they will not implement the EU-wide legislation for approvals until new, stricter, regulations for labeling and tracing GE-containing products also take effect (likely in early 2004; see "Labeling and Traceability," below).

In the 3 years before the de facto ban, U.S. corn exports to the EU averaged approximately $300 million annually (Spain and Portugal were the largest EU importers). Since then, they have declined to less than one-tenth of that value annually — the result, according to analysts, of the EU's moratorium on the approval of new corn varieties already approved in the United States. Although one variety of biotech corn was approved by the EU prior to the moratorium, other approved varieties also are being grown in the United States. Thus, U.S. exports of any corn to the EU is impractical because of the difficulty of segregating EU-approved from EU-unapproved varieties. U.S. soybean exports to the EU have declined, too, from an average value of approximately $2.2 billion in the 3 years before the ban to about $1.2 billion annually (2000-2002), according to USDA data. However, unlike biotech corn, only one variety of biotech soybeans has been widely grown in the United States; the EU had approved this GE crop under its regulatory process prior to the moratorium.

THE WTO CASE

The U.S.-led case began with a request for 60 days of formal consultations with the EU to resolve the dispute. Besides the three main filers, Australia, Chile, Colombia, El Salvador, Honduras, Mexico, New Zealand, Peru, and Uruguay expressed support for the case. On June 19, 2003, a USTR spokesman issued an announcement that consultations were not successful, and that the WTO will now be asked to establish a dispute settlement panel, the next step. Resolution of the dispute could take up to 18 months.

The United States and its allies argue that the EU moratorium violates the WTO Agreement on the Application of Sanitary and Phytosanitary (SPS) Measures. The SPS agreement permits countries to regulate crops and food products to protect health and the environment, but their rules must be scientifically justified, and approval procedures must occur without undue delay.

U.S. interests contend that there is no scientific evidence that GM-derived food and feed crops are substantially different from, or any less safe than, conventional varieties, a conclusion they say even European scientific authorities have reached. The United States contends that EU biotechnology measures also are inconsistent with provisions of the Technical Barriers to Trade Agreement, the General Agreement on Tariffs and Trade, and the Agreement on Agriculture.

The United States has argued that the moratorium has not only impacted U.S. exports to the EU, but also caused other countries — particularly in the developing world — to shun biotechnology, which the U.S. asserts holds great promise for vastly improving agricultural productivity and feeding growing populations. U.S. officials cite the recent famine in six sub-Saharan countries. A number of these countries imposed conditions on the use of GE corn food aid, and Zambia refused all GE shipments because of unspecified environmental and food safety concerns. U.S. officials argue that these actions are directly linked to the EU policy. (Some African producers may fear that the EU will refuse their agricultural exports to the EU if GE crops are widely introduced in Africa.)

EU officials counter that they have shown good faith in moving quickly to restart the approval process. In early April 2003, the EC threatened to take legal action at the European Court of Justice against the 12 member states if they fail to implement the GE approval legislation once the new labeling and tracing rules are approved. EU officials have told the United States that their cautious approach to regulating agricultural biotechnology is necessary to establish confidence among European consumers.

DIFFERING REGULATORY APPROACHES

The United States has embraced the concept of substantial equivalence with regard to a GE food or agricultural product. That is, as long as such a product is substantially the same as its conventional counterpart, it should be regulated no differently. The EU, on the other hand, takes what has been called the precautionary approach, which says that if scientific evidence is insufficient or inconclusive regarding a practice's or product's potential dangers to human or environmental health, it should be more vigorously regulated or even prohibited if there are reasonable grounds for concern, thus providing a safeguard against future unforeseen problems. Under this approach, the products of biotechnology are deemed to be inherently different than their conventional counterparts.

United States

The basic federal guidance for regulating biotechnology products is the **Coordinated Framework for Regulation of Biotechnology** (51 *Fed. Reg.* 23302) published in 1986 by the White House Office of Science and Technology Policy (OSTP). One of its key principles is that genetically engineered products should continue to be regulated according to their characteristics and unique features — not according to their method of production. Thus, if a food product produced through biotechnology is determined to be substantially equivalent to one produced by more conventional means, that food is subject to no additional (or no different) regulatory processes. Once approved, food products do not have to be labeled as to whether or not they contain any genetically modified organisms (GMOs), except to the extent a GE food is substantially different (e.g., contains an allergen or has a changed nutritional content). However, marketers are free to make such claims, one way or the other, so long as the labeling is truthful. Assuming the essential equivalence of GM products, the framework regulates new biotechnology products under existing federal statutory authorities, all of which were conceived and enacted before the advent of commercial agricultural biotechnology.[2]

European Union[3]

In contrast to the United States, the EU has established separate structures specifically for approving biotechnology crops and also for labeling products from them. Before the de facto moratorium, the EU had approved the commercial release of 18 GMOs (including "Roundup-Ready" soybeans). Four primary authorities for the EU's current GMO regulatory structure are:

---

[2] The three lead agencies are USDA's Animal and Plant Health Inspection Service, which regulates the import, interstate movement, and field testing of plant GE plants; the Food and Drug Administration, which regulates food and animal feed additives, including those derived from biotechnology, to ensure that they pose no human health risks; and the Environmental Protection Agency, which regulates pesticides, including those genetically engineered into plants. See CRS Report RL30198, *Food Biotechnology in the United States: Science, Regulation, and Issues.*

[3] Sources for this section include: USDA Office of Agricultural Affairs, U.S. Mission to the European Union, *Genetically Modified Food and Feed,* updated March 11, 2003; Pew Initiative on Food and Biotechnology, *U.S. vs. EU: An Examination of the Trade Issues Surrounding Genetically Modified Food,* 2002; and the European Commission, *Questions and Answers on the Regulation of GMOs in the EU,* March 4, 2003.

Council Directive 2001/18 (replacing Directive 90/220) spells out approval steps for assessing human health and environmental risks before any GMO or GMO-containing product can be released into the environment or marketed. The competent authority in the EU member state where the product is to be released is responsible for assessing its safety and, if approved, notifying other member states, opening the way for marketing throughout the EU. If a member state objects, the European Commission (EC) is charged with making the final decision after receiving an opinion from its Scientific Committee — although the EC decision can be approved or rejected by a qualified majority of the Council of Ministers. This new directive came into force on October 17, 2002. The directive also mandates new GMO labeling and traceability requirements. The de facto moratorium on new approvals is expected to remain in effect until these new, stricter, GMO labeling and traceability rules, now nearing final approval, take effect (see "Labeling and Traceability," below).

Regulation (EC) 258/97 on Novel Foods and Novel Foods Ingredients, administered by the Health and Consumer Protection Directorate General, governs both food safety assessments and labeling of GM foods. The safety assessment and authorization procedures are somewhat similar to those outlined above. The Novel Foods Regulation also provides for mandatory labeling of foods and food ingredients containing GMOs, which will remain in effect until the new labeling/traceability rules take effect.[4] Because the current Novel Foods Regulation lacks implementing detail, member states must determine testing methods, what to test, and so forth. No GM product has yet emerged from a full Novel Foods authorization, according to USDA.

Council Regulation 1139/98 covers foodstuffs from herbicide tolerant soybeans & Bt corn, approved before the Novel Foods rule was adopted. *Commission Regulation 50/2000* sets labeling rales for foods with GMO-based additives/flavorings.

Differing Consumer Views?

Differing U.S. and EU perspectives may reflect the fact that U.S. consumers not only are less fearful of GE foods than their European counterparts, but also are

[4] Under a separate regulation (EC) 49/2000, GM labeling has not been required if a conventional food contains less than 1% detectable GM DNA or proteins.

more confident in their food safety regulators. According to USDA's Economic Research Service (ERS), surveys of consumer attitudes toward GMOs, conducted both here and overseas, have yielded mixed results. Still, "U.S. consumers have voiced little objection to genetically modified foods, while EU consumers have been vocal in their disapproval," ERS observed. "In the United States, only a few protests have been heard and few major food manufacturers have reacted."[5]

ERS, the Pew Initiative on Food and Biotechnology, and others suggest that many Europeans may believe that food-related biotechnology is unnecessary, frivolous, or even dangerous. Europeans may be much more wary of changes in how their food is produced due to a series of recent food safety crises. During the 1990s, bovine spongiform encephalopathy (BSE, or "mad cow disease") emerged in the United Kingdom and spread to other parts of Europe. U.K. food safety authorities first insisted that the disease could not be transmitted to humans eating meat from BSE-infected animals. By 1996 scientific evidence indicated there was a link between some cases of a similar human disease and such consumption. In 1999, high levels of dioxin were found in meat products and eggs originating in Belgium. Also, foot-and-mouth disease (FMD) outbreaks in Europe added to consumer concerns and to their "waning faith in regulatory agencies," Pew concluded. "Although these crises have not been caused by GE food, GE food has been caught up in the general suspicion about food safety."[6] Vocal environment groups in the EU also have raised concerns about environmental impacts.

LABELING AND TRACEABILITY

The U.S.-led WTO case does not involve the new EU labeling regulations which soon will require all food, feed, and processed products from GMOs to be labeled, even if they no longer contain detectable traces. A tolerance level for non-GMO foods, feeds, and processed products of 0.9% is set for allowable "adventitious presence" (AP)—that is, unintended, low-level presence — of an EU-approved GE substance. All products with more than 0.9% must be labeled as GM. The allowable level for unapproved GE varieties that received a positive EU risk assessment during the moratorium is 0.5% for 3 years, after which it drops to 0%. The new EU labeling and traceability system was approved by the European

[5] ERS, *Economic Issues in Agricultural Biotechnology* (Information Bulletin No. 762), pp. 28-30. February 2001. Details on the results of these surveys can be found in this publication.

[6] Pew Initiative. Unlike BSE, scientists have not linked FMD to human health concerns, although it is a serious animal health problem.

Parliament on July 2, 2003. It is expected to come into force in early 2004 after the EU Council formally adopts it later this year.[7]

U.S. agricultural interests argue that, even if the GMO approval moratorium is lifted, the new labeling and traceability rules are themselves unworkable and unnecessary, and will continue to discriminate against U.S. exports. The U.S. grain system is competitive largely because of its high-volume, commingled nature. Compliance with the EU labeling rule would require segregation of GE crops and foods derived from them from the time they are planted all the way through the processing and marketing chain. This would entail prevention of pollen drift from GE to non-GE fields; and difficult and costly handling procedures such as using separate equipment, storage, and shipping containers, or at least painstakingly cleaning them. U.S. interests argue that food companies forced to label accurately all GE products could face huge risks and liabilities. All of these problems would discriminate against U.S. shipments, they contend.

OUTLOOK

It is unclear how a WTO dispute panel might rule on the U.S.-EU case, in part because agricultural GMOs are not explicitly recognized in the Uruguay Round trade agreements, which were concluded before the advent of widespread agricultural biotechnology. If the United States case succeeds, it would validate the basic principles of the SPS agreement and could discourage other countries from emulating the EU regulations. The United States has pointed out that many EU fanners themselves would like to be planting and selling GE crops. However, U.S. success might not open EU markets to more GE imports; the United States might simply have to settle for some form of alternate compensation. Some have suggested that a win could create a backlash among the European public and governments who view the United States as forcing biotechnology on unwilling consumers. Further, the United States is now facing the imposition of approximately $4 billion worth of WTO-sanctioned retaliatory tariffs on its exports to the EU due to its failure so far to comply with a WTO ruling to make its favorable tax treatment of U.S. Foreign Sales Corporations compliant with trade rules.

For several years, Members of Congress representing agricultural interests had urged the United States to challenge formally the EU ban, in the belief that

---

[7] For a description of the new rules, see the July 2, 2003, Commission press release at |http://europa.eu.int/rapid/start/cgi/guesten.ksh?p_action.gettxt=gt&doc=IP/03/935|0|RAPID &lg=EN&display=].

U.S. producers have been adversely impacted there, and will face further barriers if more countries take the EU approach to regulating the products of biotechnology. At the same time, many lawmakers also are well aware of the risk involved in escalating U.S.-EU trade tensions to new heights and can be expected to monitor developments closely.

The House Agriculture Committee held a hearing on March 26, 2003, on "Artificial Barriers to U.S. Trade and Food Aid," where the theme was the EU's moratorium on GMO approvals and its possible influence on decisions by developing African countries to reject or delay U.S. food aid. The Senate on May 23, 2003, passed, by unanimous consent, a resolution (S.Res. 154) supporting the U.S. action against the EU; a similar House measure (H.Res. 252) was passed on June 10, 2003, by a suspension vote, 339-80.

Chapter 8

BIOSAFETY PROTOCOL FOR GENETICALLY MODIFIED ORGANISMS: OVERVIEW[‡]

Alejandro E. Segarra and Susan R. Fletcher

INTRODUCTION

In general, "biosafety" refers to efforts to ensure safety in using, transporting, transferring, handling, releasing and disposing of biological organisms–including genetically modified organisms (GMOs)--when they are considered potentially capable of harming human, animal or plant health, or the environment. Currently, most developed countries have laws and regulations that ensure the biosafety of potentially hazardous organisms in domestic use and in trade.[1] However most developing countries lack similar legal biosafety safeguards.

The Cartagena Biosafety Protocol,[2] completed in January 2000, is the international agreement that addresses safety concerns related to trade in genetically modified organisms.[3] The Protocol applies to the transboundary

---

‡ Excerpted from CRS Report RL30594.

1 In the U.S., biotechnology is regulated jointly by the Department of Agriculture, the Environmental Protection Agency, and the Food and Drug Administration. For a thorough discussion of biotech regulation in the U.S. see CRS Report RL30198, *Food Biotechnology in the United States: Science, Regulation and Issues*, by Donna Vogt & Mickey Parish.

2 This Protocol is named after the city of Cartagena, Colombia, where the 1st Extraordinary Meeting of Conference of Parties (ExCOP) drafted significant portions of this key accord on February 14-22, 1999.

3 GMOs are organisms altered by using recombinant DNA and RNA techniques.

movement, handling and use of those GMOs that may have impacts on human health, the environment, or biological diversity. The Protocol is viewed by many as the main mechanism for dealing with biosafety concerns when GMOs cross international borders.[4] This Biosafety agreement is a protocol to the 1992 Convention for Biological Diversity (CBD–also popularly known as the Biodiversity Convention), a global agreement that addresses major aspects of biological diversity, including species conservation, genetic resources, and ecosystem protection, with a general goal of reducing species extinctions. The CBD entered into force in 1993, and at present, 179 nations are parties to this treaty.[5] The United States is not a party to the CBD (the Senate has declined to approve the treaty, as discussed below), and thus could not participate directly in the official proceedings that drafted the biosafety Protocol. However, U.S. observer delegations have attended all CBD meetings since 1992, and have played an active role in making its positions known and influencing treaty deliberations. Drafting of the Biosafety Protocol started in 1996, and in January 2000, the parties to the CBD agreed upon a final draft Protocol in Montreal, Canada.

Many, especially within the Clinton Administration, thought that U.S. actions were key in shaping the draft Protocol to protect U.S. interests, yet they acknowledged the U.S. capacity to influence CBD's deliberative bodies was affected by its non-party status. Other analysts think that the Protocol will change international trade rules for biotechnology in ways contrary to U.S. interests. They have argued that provisions in the Protocol are mostly based on controversial concepts, such as the precautionary approach (popularly known as the "precautionary principle"; see page 6), which they see as veiled threats to fair trade.

Other potentially controversial Protocol provisions, still subject to further negotiation, will require mandatory labeling of bulk commodities or could call for environmental studies, and may establish compulsory product information and disclosure procedures. In general, provisions in the Biosafety Protocol are likely to shape market rules and impact U.S. interests in biotechnology trade.

---

[4] The Biosafety Protocol refers to GMOs alternatively as "Living Modified Organisms" or LMOs.

[5] Countries become parties after signing and ratifying the CBD. Currently, 12 nations have not ratified the CBD: Afghanistan, Andorra, Bosnia-Herzegovina, Brunei, Iraq, Kuwait, Libya, Malta, Saudi Arabia, Somalia, Thailand, and the United States.

Background –The Convention on Biological Diversity

As noted above, the Biosafety Protocol is an outgrowth of the Convention on Biological Diversity. Only parties to the parent treaty, the CBD, are part of the official negotiations on protocols, which are effectively adjunct treaties on specific issues that the parties believe require elaboration or treatment. The United States played an active role in the negotiations that produced the CBD, even though the Congress has declined to approve its ratification. Similarly, because of the importance of the Biosafety Protocol to U.S. agricultural and biotechnology interests, the United States has remained actively involved, sending observer delegations, to the negotiations on it. These delegations actively participated in substantive discussions and formulation of positions that were made part of the official proceedings of the protocol negotiations. Because the United States is not a party to the CBD, it is not in a position to sign the Protocol, and thus the history and status of the Biodiversity Convention is a key element in considering the U.S. role in the Biosafety Protocol.

The Convention on Biological Diversity (CBD) dates back to intensive negotiations in the early 1990s. In 1992, world leaders came together in Rio de Janeiro, Brazil, to attend the world's largest summit –the United Nations Conference on Environment and Development (UNCED–popularly known as the "Earth Summit") to take up the wide array of issues relating to environmentally and economically sustainable use of the world's resources. One major area of concern was biological diversity – the concern that rapid loss of species around the world is occurring and international cooperation was needed to address the problem. The CBD was one of two major treaties opened for signature at UNCED (the other was on climate change).

In 1992, President Bush declined to sign the CBD treaty, citing concerns including the lack of sufficient protection for intellectual property. In 1993, President Clinton signed the CBD and sent it to the Senate for its advice and consent for ratification. The Administration accompanied the proposed treaty with a statement of interpretation in order to allay concerns raised by detractors. In 1994, the Senate Foreign Relations Committee reported favorably on ratification, but the full Senate did not act on the treaty before the end of the 103rd Congress. Since 1995, consideration of the treaty has received little or no attention from the U.S. Senate. Currently, the treaty remains pending in the Senate Committee on Foreign Relations. The CBD treaty contains few binding obligations, relying instead on largely voluntary measures. Also, the United States is in compliance with the treaty on the basis of current U.S. law, so no changes in U.S. law would be needed if it should become a party.

There are differences of opinion regarding U.S. ratification of the CBD. Some observers consider that the treaty is flawed and could be harmful to U.S. trade and national security interests. Some critics question the need for the United States to become a party to the CBD because, as the number one producer and exporter of biotechnology, the United States is already in a position of strength to set market rules. Some critics are concerned that other nations could use the Protocol provisions as disguised protectionism. Conversely, others feel that the CBD is needed to protect biological diversity, and that the United States is paying a growing price by its refusal to ratify. Backers of ratification cite a growing sense of isolation of the United States from some of its key trading partners, and they worry that the United States will have less leverage in future CBD decisions.

FRAMING THE CBD BIOSAFETY PROTOCOL

The use of biotechnology has led to broad changes in medicine, industrial processes, engineering and agriculture. In agriculture, biotechnology is being used to improve crops' resistance to pests and environmental stresses and to enhance their nutritional content. Controversies have developed among various nations over whether certain types of GMOs constitute a potential hazard to human health or to the environment. GMOs have long been considered the subject of a number of biosafety concerns. Many genetically modified plants, for example, are produced with techniques that include the DNA sequences from common plant pathogens. Other biosafety concerns commonly cited are: the possibility of adverse impacts on non-target species or ecosystems; the potential for enhanced weediness of GM crops (i.e. a plant becomes invasive, or transfers resistance genes to its wild relatives); and the instability of inserted genes (i.e. the possibility that a gene will lose its effectiveness.)[6] To date, transfer of genes to weedy relatives has been observed, and a controversial laboratory study has shown potential damage to a non-target species (the monarch butterfly).[7] Whether such effects prove to be widespread or manageable has not been fully established.

---

[6] The consensus among U.S. scientists is that no strict distinction exists between the health and environmental risks posed by GM plants and those modified by conventional breeding practices. Opponents counter that risks are potentially irreversible, untraceable and uncontrollable.

[7] Metz, et al. 1997. The impact on biosafety of phosphinothricin-tolerance transgene in interspecific B. rapa x B. napus hybrids and their successive backcrosses. Theor. Appl. Genet. 95:442-450. Springer-Verlag (weeds).; and Losey, J., et al., 1999. Transgenic pollen harms monarch larvae. Nature 399: 214.

Biosafety Protocol Negotiations

The CBD's Biosafety Working Group (BSWG) negotiated the Biosafety Protocol in six meetings that took place between 1996 and January of 2000. After the Earth Summit in Rio, the BSWG became the principal forum for deliberations and negotiations for biosafety issues, regarded by parties to the CBD as an important area of "unfinished business."

According to observers, one difficulty in these negotiations was the fact that the United States, as the largest exporter and producer of GMOs, was not officially part of the proceedings. However, the United States actively participated as part of the "Miami group," a coalition of leading agricultural exporters that also included CBD parties: Argentina, Australia, Canada, Chile and Uruguay. This group developed proposals and positions during the drafting of the Protocol.[8]

According to the CBD Secretariat, the Protocol will enter into force by August 2000, having met the criterion in Article 37 of the Protocol at the closing of the 5th Meeting of the Conference of the Parties (COP5) in Nairobi, Kenya on May 26, 2000.[9] For many parties, signature to a protocol automatically confers ratification, since they have ratified the parent treaty; in others, like the United States, each protocol must be separately considered for ratification after signature. The Protocol is currently open for signatures by parties and will remain open at the United Nations Headquarters until June 5, 2001. Nations that wish to sign onto the Protocol after that date would have to do so by accession, which includes ratification.

Analysts agree that, while the Protocol is a substantially complete document, the pact delays final negotiations on a number of controversial issues for up to 4 years after its adoption. For example, the door has been left open for parties to revise or strengthen rules for 'contained use' GMOs (discussed below) that are not likely to propagate in the environment (i.e., bulk commodities.) The Miami group is expected to oppose any move to increase documentation and notification requirements for bulk commodities, arguing that the cost of segregation and identity preservation (i.e., tracing a GMO product from the farm to the consumer's table) could lead to irreparable harm to biotechnology trade.

---

[8] Other key interest groups that formed and were active during drafting of the Protocol were the European Union (EU), the "Like-minded" group of developing countries (the G-77/China, less the members of the Miami group), and the "Compromise group," (Japan, Mexico, Norway, the Republic of Korea and Switzerland.)

[9] Article 37 specifies that the Protocol will enter into force on the 90th day after the date of deposit of the fiftieth instrument of ratification. As of December 22, 2000, two countries (Bulgaria and Trinidad Tobago) have ratified the protocol and 79 others have signed it.

Several important aspects of current provisions have been left for future negotiations such as: (1) how to apply the "precautionary approach" to the Protocol; (2) how to develop rules to integrate the Protocol with other trade agreements, such as the WTO, into a "mutually supportive" framework; and (3) how to develop rules and procedures based on international law that would establish liability and redress for damages resulting from the movement of GMO (these are to be finalized within 4 years).

The Biosafety Protocol: Key Provisions and Related Issues

The Protocol addresses the major concerns that have arisen in connection with GMOs, and it bears directly on GMO trade. Most GMOs are covered by the Protocol except for those used to produce human pharmaceuticals. The key provisions, and the issues related to them, are briefly summarized below.

Advance Informed Agreements

The Protocol establishes the use of 'Advance Informed Agreements' (AIA) between the importing and exporting parties that cover the first transboundary movement of any GMO. The purpose of AIAs is to ensure that recipient countries have the opportunity to assess environmental risks associated with the importation of biotechnology products. The Biosafety Protocol creates the procedure by requiring exporters to seek consent from importers before *the first shipment* of a GMO is introduced into the environment (applies to seeds for planting, fish for field release, and microorganisms for environmental bioremediation). In addition, article 11 also requires that bulk shipments of GMO commodities to be used as food, feed, or for processing must be accompanied by declarations stating that such shipments "*May Contain*" GMOs and are "*Not intended for intentional introduction into the environment.*"

Two categories of GMO are recognized according to their intended use: GMOs for 'contained' or 'direct' use (i.e., food, feed, or for processing) that require minimal biosafety precautions; and GMOs for 'intentional introduction' to the environment (i.e., agricultural seeds and other propagation materials, and live fish), which require more stringent biosafety procedures. GMOs unlikely to cause adverse effects on biological diversity, and those intended for contained use (i.e., feed, food, or processing) can be *exempted* from the application of AIA procedures. Confidential information received under Protocol procedures must be protected by importing and exporting parties. Further, parties are required to cooperate in developing and/or strengthening the biosafety capabilities of developing countries.

Key Issues

The rationale for Advance Informed Agreements (AIA), espoused in 1992 by article 19.3 of the CBD, is the perception that international trade in GMOs could adversely affect the environment and that trade should not proceed without the informed agreement of the recipient country. While support for some form of AIA is widespread, differing positions about their scope and breadth were among the principal issues under discussion. For example, as negotiations progressed it became clear that the GMO's intended use should determine the level of risk assessment and risk management needed to attain biosafety. Similarly, negotiators decided that the intended use of a GMO should determine the level of information disclosure needed from the exporting party. However, the issue of labeling and proper documentation for bulk commodities will be revisited by the Conference of Parties (COP) within 2 years, thus opening the door for the introduction of additional restrictions and mandatory requirements.

Precautionary Approach

The Protocol reaffirms the importance of the precautionary approach in its preamble and in articles 10 and 11, which prescribe trading procedures for the importing parties. These articles, for instance, state that even in the case of a lack of scientific certainty (due to insufficient scientific data, analysis or information) importers could deny entry to undesired GMOs. At the same time, risk assessments and risk management are prominent features of the Protocol. The Protocol allows the importer to require that the exporter carry out and pay for risk assessments as well as any needed mitigation actions. In addition, the Protocol allows parties to use socioeconomic considerations in reaching a decision on importing GMOs.

The "precautionary approach" was established as a guiding tenet of the CBD, which directly referred to Principle 15 of the non-binding Rio Declaration. Principle 15 states that:

> "In order to protect the environment, the precautionary approach shall be widely applied by States according to their capabilities. Where there are threats of serious or irreversible damage, lack of full scientific certainty shall not be used as a reason for postponing cost-effective measures to prevent environmental degradation."

Key Issues

In general, most interpretations agree that the precautionary approach urges policymakers to err on the side of caution in the face of scientific uncertainty.[10] Environmentalists, particularly in Europe, consider this approach to be a valid policy option in the face of incomplete or inadequate scientific knowledge about health or environmental impacts. Defenders of the application of this approach maintain that it is only "a temporary mechanism" that gives time for scientific inquiry. Policymakers in Europe recognize, as do those in the United States, the need for an assessment of risks based on accepted scientific facts.[11] However, critics worry that elevating the precautionary approach to the level of a 'political principle' may create false public expectations for absolute safety and the demand for zero environmental risks. At worst, some critics maintain, the precautionary approach can easily be used as a form of disguised protectionism.

Current language in the Protocol leaves the development of procedures for trade and/or entry decisions subject to further refinement by the CBD's Conference of Parties. Presently, few countries agree on just what the precautionary approach is. Analysts believe that future refinements may be affected by this lack of accepted disciplines to guide the uniform application of the precautionary approach. The current situation has enticed international bodies, like the EU, the Codex Alimentarius Commission, and the Organization for Economic Cooperation and Development (OECD), to offer comprehensive views on the subject in hopes of setting a world's standard for the application of the principle in GMO trade.[12]

Biosafety Clearinghouse

The Protocol establishes the Biosafety Clearing-House (BCH), which will be used as the mechanism to share scientific, environmental, and legal information on GMOs.

---

[10] For a discussion on the precautionary approach see CRS Report RS20310, Science behind the regulation of food safety: risk assessment and the precautionary principle by Mickey Parish.

[11] Internal political disagreements exist within the EU over the use of the precautionary approach. E.g, France is at odds with the EU over France's use of precautionary approach for not admitting British beef into the country after the "mad cow" disease.

[12] OECD deliberations on the precautionary approach can be found at http://www.oecd.org/subject/biotech/edinburgh.htm . The European Union's position is presented in "*Communications from the Commission on the precautionary principle*" February 2, 2000. Brussels. (see http://europa.eu.int/comm/trade/whats_new/dpp_en.htm.); Codex Alimentarius position found in "*Report of the First Session of the Codex Ad Hoc Intergovernmental Task Force on Foods Derived from Biotechnology*", April 2000, in http://www.fao.org/waicent/faoinfo/ECONOMIC/esn/codex/default.htm.

As currently envisioned, the BCH will be a centralized source for information on national laws, guidelines and regulations for the implementation of the Protocol, as well as on other agreements on biotechnology. The need for a centralized information source on scientific and technical data about GMO trade was also envisioned in 1992 by the CBD (Articles 18.3 and 19.) To accomplish this, Article 20 of the Protocol establishes an internet-based "Biosafety Clearing-House" to help countries exchange information about GMOs. Currently, the BCH is housed at the CBD's Executive Secretariat in Montreal, Canada.

The Protocol requires that transboundary movements of GMOs between parties and non-parties be consistent with the Protocol and encourages non-parties to contribute information through the BCH. Parties are required to notify the BCH of unintended or illegal movements of GMOs, and must adopt measures to prevent and penalize illegal transboundary movements.

Key Issues

Two critical issues were foremost in the minds of negotiators in establishing the BCH: (1) how to adequately balance the rights of parties to access and use relevant information about traded GMOs; and (2) how to protect the intellectual property rights of owners of biotechnology products. Article 20(3) and Article 21 address the protection of confidential and proprietary information by making both trading parties responsible for protecting confidential information received under the Protocol under AIA requirements. However, some controversy exists because the Protocol does not prescribe explicit liabilities for failures to protect intellectual property. Some critics would like to see clear-cut mechanisms in the Protocol to sanction failures to protect confidential information. Recent proposals by by Canada to establish a pilot implementation phase and to have each country apply its domestic laws and requirements to information sharing seem to be favored by other Miami group members.[13] Formal procedures and measures of intellectual property protection are to be developed by the Conference of Parties within 2 years.

Relationship of the Protocol with other Agreements: The Savings Clause

One of the most contentious issues faced by negotiators was to establish how the Protocol's measures would relate to other bilateral or international trade agreements, notably those under the World Trade Organization (WTO). The issue became important after the 1997 Biosafety Working Group meeting in Montreal,

[13] First Intergovernmental Committee for the Cartagena Protocol on Biosafety (ICCP) December 11-15, 2000 . Montpellier, France.

where the Miami group insisted on the need for consistency between the Protocol and WTO agreements such as the Sanitary Phytosanitary-Standards (SPS) accord.[14] The key question revolved around how to reconcile differing perspectives in environmental protection philosophies between advocates of the precautionary approach (e.g., the EU), and countries where environmental protection systems stress the use of the best available scientific evidence and risk assessment (i.e., the Miami Group and the United States).

A compromise position was reached through the inclusion of a "savings clause" in the Protocol's preamble that states: "*...nothing in the Protocol implies a change in the rights and obligations of governments under the WTO or other existing international agreements.*" However, the preamble further clarifies that the Biosafety Protocol will not be subordinated to other international agreements, and that they should be mutually supportive.

Key Issues

The Clinton Administration and industry analysts interpret the 'savings clause' to mean that trade disputes originating from Protocol implementation could be handled through existing mechanisms, such as the WTO dispute settlement agreements and its appellate review bodies. Others, especially EU policymakers, are inclined to stress that the Protocol is not subordinated to any agreement, and do not necessarily extend the 'savings clause' to future agreements or to existing ones, which do not adhere to the Protocol.

Notification/Labeling

The Protocol establishes mandatory entry notifications by exporting countries to the competent national authority in importing countries about trade of non-exempted GMOs. A key feature of the Protocol is the requirement for bulk shipments of GMO commodities to be accompanied by documentation stating that such shipments "*May Contain*" GMOs, and that they are "*Not intended for intentional introduction into the environment,*" in lieu of a formal AIA. This applies only to GMOs intended for food, feed, or processing (e.g., corn or soybeans).

Other Provisions

Finally, parties are allowed to enter into bilateral, regional or multilateral agreements provided that such agreements and arrangements do not result in a

[14] This accord establishes disciplines on the use of sanitary and phytosanitary measures in restricting trade. For a complete discussion on SPS see CRS Report 98-254 ENR. *Agricultural Negotiations in the World Trade Organization* by Charles E. Hanrahan.

lower level of protection than that provided by the Protocol. Also, in the case of unintended or illegal movements, the Protocol also allows the affected party to request the party of origin to dispose of the GMO in question at the expense of the latter. In addition, the Protocol calls for the elaboration and adoption of a liability and redress process for damages resulting from transboundary movements of GMOs.

IMPACTS ON TRADE

Analysts agree that the future impact of the Protocol on trade is difficult to gauge because so many key aspects are still to be decided. Also, the current climate of controversy surrounding trade in agricultural GMOs in many countries, especially in Europe, further complicates assessments of trade implications. The Clinton Administration and the U.S. biotechnology industry have stated that the new rules will make it easier to harness the promise of biotechnology without unduly disrupting world food trade.[15] Others, especially critics in Congress, have viewed the accord with skepticism, and have charged that many of its provisions will harm biotechnology trade by opening a potential "flood gate" to restrictive and costly labeling and documentation requirements for U.S. goods.

CONCLUSION

The United States has pioneered the use of biotechnology and is the world's leading producer of GMO crops. While most U.S. analysts consider that the current Protocol language is a victory for GMO trade, others point to likely future battles and continued friction between the United States and Protocol parties over unresolved areas. Some, particularly in Congress, felt that the Clinton Administration had gone too far in giving concessions for the inclusion of the precautionary approach, and criticize U.S. negotiators for placing the future of GMO trade on a slippery slope by allowing even a modicum of labeling requirements in bulk commodities. Others within that Administration defended their efforts for keeping human health issues out of the Protocol; for lobbying to protect existing trade obligations by introducing the 'savings clause'; and for

[15] The United States, through the Miami Group, was a key lobbyist for the "savings clause"and for securing minimal AIA procedures for bulk shipments of GMO foods and feed. The savings clause establishes the relationship between the Protocol and other international trade agreements.

insisting upon increased intellectual property protection guarantees. They saw these efforts as victories fought and won under the handicap of U.S. non-party status within the CBD. Differences in viewpoints on ratification persist, with some who feel that ratification of the CBD would give the United States a seat at the table and the power to make its full weight be felt, and others who feel that the United States can continue to have adequate influence on the process without being party to a treaty that they consider to be flawed.

These issues related to GMOs and how the Biosafety Protocol may affect U.S. interests related to them are likely to be of continuing interest to Congress in terms of possible economic impacts and trade considerations.

Chapter 9

AGRICULTURAL BIOTECHNOLOGY*

Geoffrey S. Becker

Since the first genetically engineered (GE) crop varieties became commercially available in the mid-1990s, U.S. farmers have been rapidly adopting agricultural biotechnology in an effort to improve on-farm productivity, profitability, and competitiveness. In 2001, nearly 70% of all U.S. cotton and soybean acreage, and about 25% of all U.S. corn acreage, were planted with GE seed varieties, designed mainly to control pests. The next generation of GE commodities could shift the focus of agricultural biotechnology from the input side (i.e., farm production benefits) to the output side (i.e., consumer benefits). GE farm products also may find widespread pharmaceutical and industrial uses.

POLICY ISSUES

Many farmers are looking for lower production costs, higher yields, and new market opportunities through use of GE crop varieties. However, introduction of the technology also has been controversial. Concerns were heightened in 2000 when a Bt variety of corn (Starlink) not approved for human food use was detected in taco shells. (In December 2002, traces of Starlink were detected in a shipment of U.S. corn in Japan, where it is prohibited.) In 2002, GE-altered corn

* Excerpted from CRS Report BA-7008. Updated Jan. 9, 2003.

plants being test-grown in Nebraska for pharmaceutical use (for ProdiGene, Inc.) were inadvertently mixed with some 500,000 bushels of soybeans, which subsequently were quarantined by USDA to keep them out of the food supply. Public policy issues include:

- The impacts of GE crops on the environment and biodiversity, and on food safety and consumer acceptance, including the controversial issue of GE labeling;
- The extent to which a few large companies are gaining control over GE seed varieties - and how that might affect their cost and availability, the agricultural research agenda, patent ownership rights, and so forth;
- Impacts on U.S. fanners and the agricultural production system;
- Whether current U.S. regulation and oversight of biotechnology are adequate, particularly as newer applications emerge;
- Global trade implications, as some major U.S. agricultural export destinations, notably the European Union (EU), take a much more restrictive approach to regulating agricultural biotechnology;
- The prospects for agricultural biotechnology to improve (according to proponents or undermine (according to opponents) food security in developing countries and elsewhere in the world.

Adoption of Bioengineered Crops

Approximately 130 million acres of various GE crops are now planted worldwide, more than two-thirds of them in the United States, according to USDA. Generally, GE crops in the United States are aimed at more efficient pest control.

1. Herbicide tolerant (HT) crops are engineered to tolerate herbicides that would otherwise kill them along with the targeted weeds. These include so-called "Roundup Ready" or HT soybeans, HT cotton, and to a lesser extent, HT corn.

2. Insect-resistant crops effectively have the pesticide inserted into the plants themselves to control insect pests for the life of the crop. Currently Bt *(Bacillus thuringiensis,* a soil bacterium) crops are the only ones in commercial use. They are most prevalent in cotton to control tobacco budworm, bollworm, and pink bollworm, and in corn to control earworm and several types of corn borers.

U.S. Adoption of Major Bioengineered Crops, 1996 and 2002 (Percent of Total Acreage of each Crop)

	HT soybeans	HT cotton	Bt Corn	Bt Cotton
1996	7.4	2.2	1.4	14.6
2002	75	58	24	35

USDA, NASS. Crop Production (Supplement): Acreage. various issues.

FEDERAL REGULATORY ROLE

The basic federal guidance for regulating biotechnology products is **the Coordinated Framework for Regulation of Biotechnology** (51 *Fed. Reg.* 23302) published in 1986 by the White House Office of Science and Technology Policy. One of its key principles is that genetically engineered products should continue to be regulated according to their characteristics and unique features - not according to their method of production.

Thus, if a food product produced through biotechnology is substantially equivalent to one produced by more conventional means, that food is subject to no additional (or no different) regulatory processes. The framework maintains that new biotechnology products are regulated under existing federal statutory authorities and regulation. The three lead agencies and their roles are:

- **The Food and Drug Administration,** which regulates food and animal feed additives, including those derived from biotechnology, primarily to ensure that they pose no human health risks.
- **The Environmental Protection Agency,** which must approve the use of all pesticides, including those genetically engineered into plants.
- USDA's **Animal and Plant Health Inspection Service,** which regulates the importation, movement, and field testing of both plant and animal GE products.

The U.S. policy framework contrasts with that of many major trading partners: the EU, Japan, South Korea, China, Australia, and New Zealand either have or are establishing separate mandatory labeling requirements for products containing genetically modified ingredients. These differing regulatory approaches (where consumer and some official attitudes toward GE foods are more skeptical) have begun to disrupt U.S. exports and heighten trade tensions.

IN CONGRESS

In the 107th Congress, lawmakers included, in the 2002 farm bill (P.L. 107-171), provisions aimed at supporting use of GE farm products. Section 3204 (7 USC 5679) creates a "Biotechnology and Agricultural Trade Program" and authorizes appropriations of up to $6 million annually for technical assistance and public and private grants to remove or mitigate significant foreign regulatory nontariff barriers to U.S. exports involving agricultural commodities, including those produced through biotechnology. Section 7210 (7 USC 5921) authorizes appropriations for competitive grants for biotechnology risk assessment research; section 7505, (7 USC 7621) authorizes appropriations for agricultural biotechnology research and development for developing countries; and section 10802 (7 USC 592la) authorizes a science-based public education program on the use of biotechnology for producing human food. Separate legislation (PL.107-210) providing for trade promotion ("fast track") authority calls on U.S. officials negotiating trade agreements to press for the elimination of unjustified restrictions, including labeling, for biotechnology products.

Generally, the 108th Congress is likely to continue to support biotechnology in U.S. agriculture. Since controversy over impacts and benefits is expected to persist, Congress also will continue close oversight of the issue.

Chapter 10

STATUS OF GENETICALLY ENGINEERED WHEAT IN NORTH AMERICA[*]

Gwenell L. Waters Bass

INTRODUCTION

Genetically engineered (GE) wheat varieties developed by the biotech industry and university researchers hold considerable promise for producers. GE wheat developed by chemical company Monsanto is engineered to resist damage from the widely used Roundup Ready (RR) herbicide, making it easier for farmers to control weeds. The development of GE wheat raises issues of market acceptance and agronomic management Market acceptance of GE wheat focuses on its direct acceptance by consumers, either segregated or when co-mingled with non-GE or conventional wheat Most research suggests GE foods are safe to eat; however, uncertainty about health and environmental effects has led to public opposition particularly in Europe and Japan. This is of great concern to U.S. wheat producers because the United States exports more than half of its wheat production. Producers are concerned as to whether or not a segregation system can be designed which ensures that no co-mingling between GE and non-GE wheat will occur. Agronomic management issues include the effect of GE wheat on crop management practices and profitability, issues of contamination and

---

[*] Excerpted from CRS Report RS21620. September 11, 2003

spread, the development of pesticide resistance, and the cost and management of volunteer plants. While developers of the technologies are seeking regulatory approval, they indicate that they will postpone their commercialization until market acceptance is assured and systems for separating GE and non-GE wheat are in place.

BACKGROUND

GE wheat was first produced at the University of Florida in 1992. The strain of wheat that was developed was herbicide-tolerant The University holds a patent on the GE wheat technology, but has granted an exclusive license to the chemical and seed company Monsanto.[1] Herbicide-tolerant crops were developed to survive certain herbicides that destroy a crop along with the targeted weeds. With herbicide-tolerant crops, farmers can use potent postemergent herbicides, providing a more effective weed control than otherwise. Monsanto's GE wheat is a Roundup Ready (RR) crop that is resistant to glyphosate, a herbicide effective on many species of grasses, broadleaf and other weeds.

Since the first GE crops (mainly corn, soybeans, and cotton) became commercially available in the mid-1990s, U.S. farmers have been rapidly adopting them.[2] Proponents of agricultural biotechnology say it offers the potential to increase crop production, lower fanning costs, improve food quality and safety, and enhance environmental quality. Cited advantages of RR products are that they cut herbicide costs and make spraying and cleaning up of fields more convenient. For example, a key aspect of RR soybean is that it provides enormous gains in crop and time management flexibility, freeing up labor time for other activities. Thus, even if output remained the same, cost would be lower for weed control chemical applications and mechanical tillage. Since producers pay more to seed companies for the herbicide-tolerant seeds, the profitability of the herbicide-tolerant varieties depends on weed control cost savings compared with seed cost premiums. Presently, the United States has approved 12 different plants with GE traits for commercial use. Two of these plants which are widely used are com and soybeans. A study by the United States Department of Agriculture (USDA) found the adoption of herbicide-tolerant corn improved net farm returns of specialized

---

[1] Monsanto is the third largest agri-chemical company worldwide. It is the only company besides Sygenta (second largest) that has developed a potential GE wheat product.

[2] For background see CRS Report RS21381, *Adoption of Genetically Modified Agricultural Products*.

corn farms. Herbicide-tolerant soybeans, however, did not have a significant impact on net farm returns in either 1997 or 1998.[3]

Leonard Gianessi, a senior research associate with the National Center for Food and Agriculture Policy in Washington, said the GE wheat could solve a key problem for producers experiencing relatively low wheat prices. In some areas, according to Gianessi, current weed-control measures cost about $30 an acre. The use of GE wheat could reduce costs to $16 an acre. Monsanto's GE wheat could make it affordable to plant more according to Gianessi. These estimates do not include the potential cost of separating GE from non-GE wheat.[4]

The Canadian National Farmers Union of (CNFU), is a group that is opposed to the introduction of GE wheat. CNFU has cited several reasons for its opposition of GE wheat. The list includes: food safety, price impact, market loss, health concerns, environmental damage and agronomic costs. One agronomist estimates that costs to control (unwelcome present) GE RR wheat in GE RR canola (and vice versa) will require additional chemical applications (non-RR) totaling up to$400 million annually.[5]

IMPLICATIONS FOR U.S. WHEAT-IMPORTANCE OF WHEAT TRADE

The commercial acceptance of GE wheat is an issue of major importance for U.S. wheat farmers because it depends on the attitudes of consumers, domestic and foreign, and on the expected impact on food safety, health, and the environment. The United States is the fourth largest wheat producing country, with its output exceeded only by the EU, China, and India. In 2002, wheat ranked fourth among U.S. field crops in planted acreage, behind soybeans, corn, and hay. During that same period, wheat represented 6% of gross cash farm receipts, third highest among U.S. field crops.

The United States is the world's largest wheat exporter, with wheat accounting for approximately 7.5 % of the nation's agricultural export value in 2002. U.S. wheat exports equal almost one-fourth of total world wheat exports. Although nearly half of U.S. wheat is sold to foreign customers, U.S. wheat exports have shown little to no increases since 1996/97. Global competition has

---

[3] United States Department of Agriculture. Economic Research Service. *Adoption of Bioengineered Crops*. AER-810. 68pp. May 2002.

[4] Melcer, Rachel. "Coalition Asks For In-Depth Study of Genetically Modified Wheat." *St. Louis - Dispatch*. Mar 12, 2003.

[5] http://www.nfu.org

intensified in recent years as several non-traditional wheat exporters have emerged (Russia, Ukraine, India in particular) and the United States has seen its market share decline. The degree of foreign acceptance can significantly affect international trade and may create the need to identify and segregate GE wheat.

Market loss is a major concern of U.S. wheat producers because some of the United States' biggest trading partners have raised issues about accepting GE wheat if it is commercially traded. Approximately 70% of the Upper Midwest's hard red spring wheat, for example, is sold to the EU and Japan, where consumers shun biotech crops like GE corn and soybeans. Resistance to GE wheat may be even greater than for corn and soybeans, which are principally animal feeds, since wheat is eaten by humans directly. U.S. corn exports to the EU have dropped from approximately $300 million in the mid-1990s to less than $10 million in recent years. This decline is due primarily to the fact that new biotech corn varieties have not been approved for marketing in the EU.[6]

Regulatory Status of GE Wheat

United States

In December 2002, Monsanto submitted a petition to the USDA and Canadian authorities for regulatory approval of its GE spring wheat.[7] USDA's Animal and Plant Health Inspection Service (APHIS)conducts programs to protect American agriculture against pests and diseases. Generally, before a genetically engineered crop can be produced on a wider scale and sold commercially, its creators must petition APHIS for a "determination of non-regulated status." APHIS must be provided scientific details about the genetics of the plant, the nature and origin of the genetic material used, information about indirect effects on other plants, field test reports, and information unfavorable to the petition. All petitions are published in the *Federal Register* for public comment. APHIS grants the petition only if it determines that the plant poses no significant risk to other plants in the environment.

The EPA, which would have to set a tolerance for glysophate for the wheat variety, is waiting for the USDA to complete its work. Monsanto then may

---

[6] For background see CRS Report RS21556 *Agricultural Biotechnology: The U.S.-EU Dispute.*

[7] Three agencies using a coordinated framework are primarily responsible for regulating biotechnology in the United States - the USDA, the Environmental Protection Agency (EPA), and the Food and Drug Administration (FDA). For background see CRS Report RL21556 *Food Biotechnology in the United States: Science, Regulation, and Issues.*

approach the FDA for a premarket consultation, in which the company supplies its information indicating that the variety is safe to eat, but this consultation is not required by law. While Monsanto expects that its wheat will pass the long regulatory process, company officials have pledged not to sell the GE wheat seeds to farmers until the company can guarantee domestic and export markets. They also promised to help develop grain-handling and testing procedures that can be used to separate its modified wheat from non-modified wheat.[8]

On March 9, 2003, a coalition of environmental and farming groups filed a petition asking the USDA to conduct in-depth environmental and socioeconomic impact studies before considering approval for GE wheat products. The groups consist of the Center for Food Safety based in Washington DC, Dakota Resource Council based in North Dakota, Northern Plains Resource Council based in Montana, and other farm groups. The Environmental Working Group and Greenpeace have also questioned commercialization of GE crops.[9]

Their petition to the USDA sought three actions: (1) institute a moratorium on GE wheat until all possible environmental, human health and socioeconomic impacts are analyzed; (2) institute a moratorium until possible impacts under the Migratory Bird Treaty Act are assessed; and (3) classify GE wheat as a noxious weed and thereby prohibit its production.

On April 25, 2003, the National Association of Wheat Growers, U.S. Wheat Associates, and the Wheat Export Trade Education Committee sent a letter to Agriculture Secretary Ann Veneman in response to petitions. The three national farm groups represent most U.S. wheat fanners. The organizations, while acknowledging that there are market acceptance concerns to be addressed, say that a science-based regulatory system is not the appropriate place to address non-scientific concerns. They suggest that a mechanism other than the regulatory process must be found to address the timing of commercialization, so the scientific safety determinations may proceed free of non-scientific encumbrances.

A statement issued jointly by the three national organizations said that wheat producers would work with all segments of the industry to develop and assure that a viable identity preservation system and testing program would be instituted prior to commercialization of GE wheat varieties. They urged technology providers to obtain international regulatory approval and to ensure customer acceptance prior to commercialization. The wheat organizations declared that they support voluntary labeling of food products, but oppose government-mandated labeling of

---

[8] Monsanto. *Bringing New Technologies to Wheat: Information on the Development of Roundup Ready Wheat.* Pamphlet.

[9] National Association of Wheat Growers, U.S. Wheat Associates, and Wheat Export Trade Education Committee letter to Agriculture Secretary Ann Veneman. April 25, 2003.

wheat products in both U.S. and international markets, based upon the presence or absence of biotechnologically derived traits that do not differ significantly from their conventional counterparts.

In their letter responding to the March 9th petition, the three national wheat organizations argue that the use of Roundup Ready wheat treated with glyphosate as an alternative to other types of herbicide treatment will have virtually no impact on migratory birds. They also note that herbicide resistant weeds are nothing new in agriculture and have been a problem long before the advent of biotechnology.

Canada

Producers in Canada are also concerned about the acceptance of GE wheat in export markets. Canada exports more than 80% of its wheat to Europe and Japan. In Canada, the Canadian Wheat Board (CWB), which enjoys a monopoly on sales of wheat to export markets, has developed a set of conditions it says must be met before a GE wheat variety could be released. The set of conditions includes four areas: market acceptance, segregation systems, agronomic information, and cost-benefit analysis.

The CWB's condition on market acceptance is that there must be identified markets for the entire production of GE or co-mingled wheat for multiple years. There must also be the ability to meet requirements in non-GE wheat markets, including the establishment of achievable tolerance levels for GE wheat in non-GE wheat shipments. On segregation, the CWB maintains that in order to maintain access to non-GE wheat markets, an effective segregation system must be developed. This would entail not only systems to ensure that the entire production goes to identified GE wheat markets, but also that non-GE wheat shipments would be segregated and monitored from farmer to consumer. On farm management issues, the CWB framed a series of questions to be answered before the introduction of GE wheat varieties. Questions concern the effect of introduction of GE wheat varieties on management practices and profitability with respect to different kinds of fanning operations (e.g., conventional tillage, conservation tillage, organic, pesticide-free, etc.) across a multi-year rotation. The CWB called for research on other agronomic risks such as contamination and spread of genetic material, the development of weed-resistance, and the cost and management of volunteer plants.

As to cost-benefit analysis, the CWB called for a comprehensive analysis to account for the balancing of risks and costs associated with GE-wheat. Such an analysis would incorporate possible market benefits and costs, including segregation costs, as well as agronomic benefits and costs. The CWB suggested that in situations where there is some uncertainty as to future benefits and costs associated with an irreversible decision, there is always some benefit to waiting.[10]

On May 27, 2003, the CWB asked Monsanto Canada to withdraw its application for its environmental safety assessment of its GE RR wheat product to protect wheat exports to key foreign markets. The CWB asked Monsanto to confirm its compliance with the request to withdraw its application by June 27, 2003. The CWB was concerned that the unconfined release of GE wheat in Canada would result in significant and predictable economic harm to western Canadian farmers. Monsanto did not withdraw its application.

A recent study prepared for the CWB by three agricultural scientists concluded that the unconfined release of GE wheat into western Canada would threaten the environment and be environmentally unsafe.[11] The report points to earlier experiences in Canada with GE canola and claims that if glyphosate-resistant wheat were granted unconfined release in western Canada, "the trait would move between wheat cultivars and fields in a fashion similar to that seen in canola."

Other Countries

Internationally, regulations being developed with respect to genetically modified crops could lead to potential loss of markets for producers and the additional marketing costs of segregating and/or identity-preserving. These phenomena, in turn might reduce incentives for the development of new agricultural biotechnology products. Both Canada and the United States have approved more than thirty-five GE products and allow voluntary labeling of them. The EU, on the other hand, has maintained a moratorium on genetically modified products since 1998. Prior to that the EU approved ten products and it requires mandatory labeling of GE products. While Japan has approved at least twenty products, but also requires mandatory labeling.

On May 13, 2003 the United States, Canada, and Argentina initiated a case with the World Trade Organization (WTO) against the EU over its five-year

[10] Grain Industry Working Group on Genetically Modified Wheat. *Conditions for the Introduction of Genetically Modified Wheat.* Draft February 5, 2003.

[11] R.C. Van Acker, A.L. Brute-Babel, and L.F. Friesen. *An Environmental Safety Assessment of Roundup Ready Wheat: Risks for Direct Seeding Systems in Western Canada.* University of Manitoba: June, 2003.

moratorium on approving new agricultural GE products. The WTO case will go forward in spite of the EU labeling agreement. On July 2, 2003 the EU reached a compromise agreement that settles differences over pending GE products traceability and labeling regulations. The new traceability and labeling laws will effectively require U.S. farmers to separate GE crops from non-GE crops if they want to export to the EU. Approval of the traceability and labeling legislation has been one of the conditions imposed by some EU member states for allowing a renewal of GE authorizations. The threshold level for labeling a GE product is 0.9 %. In the absence of a system to segregate biotech from non-biotech products all U.S. and Canadian wheat would have to be labeled as containing or possibly containing GE products.

CONCLUSION

The potential benefits of GE wheat are great, according to proponents, but some producers foresee many risks as well. Market acceptance is the biggest concern especially in export markets. Strict segregation of GE and other wheat for export may be required in the field and during storage, transportation and processing. Segregation could raise production costs for all growers, not just those who produce GE varieties. Full-scale commercialization of GE wheat appears to be many years away and will depend as much on consumers' acceptance as on government regulation.

Chapter 11

GENETICALLY ENGINEERED SOYBEANS: ACCEPTANCE AND INTELLECTUAL PROPERTY RIGHTS ISSUES IN SOUTH AMERICA[†]

Randy Schnepf

SUMMARY

U.S. soybean growers and trade officials charge that Argentina and Brazil — the United States' two major export competitors in international soybean markets — gain an unfair trade advantage by routinely saving genetically engineered (GE), Roundup Ready (RR) soybean seeds from the previous harvest (a practice prohibited in the United States). These groups also argue that South American farmers pay no royalty fees on the saved seed, unlike U.S. farmers, who are subject to a technology fee. The cost saving to South American soybean growers on the technology fee alone nets out to about $8 to $9 per metric ton — a considerable cost advantage over U.S. soybeans in the highly competitive international soybean market. This practice also raises concerns about the intellectual property rights (IPR) of Monsanto (the developer of RR technology).

Commercial use of RR soybeans in Brazil remains illegal despite apparent widespread planting. A 1998 government approval of their commercial use remains suspended by court injunction, and resolution over their commercial

[†] Excerpted from CRS Report RS21558. July 1, 2003.

legality is being considered by an appellate court. The eventual outcome on commercial legalization of GE crops in Brazil may have important consequences for intellectual property rights, as well as for international trade in GE crops.

INTRODUCTION

The United States is the world's leading producer and exporter of soybeans. However, South American soybean production and trade has expanded rapidly during the past 15 years, greatly increasing the competitiveness of international oilseed markets. Together, the United States, Argentina, and Brazil are expected to produce nearly 83% of the world's soybeans in 2002/03, and to account for over 90% of all soybeans traded on international markets.[1] Given a highly competitive international soybean market and growing international debate over the nature of production and trade in genetically engineered (GE) crops, controversy has emerged in recent years over the growing pirated use of Roundup Ready (RR) soybeans, a GE variety, by producers in Argentina and Brazil.[2] This practice appears to provide a competitive advantage to Argentine and Brazilian soybean exports, and to be a violation of the intellectual property rights (IPR) of the RR technology producer, Monsanto.

ROUNDUP READY SOYBEANS

Roundup Ready (RR) soybeans are genetically engineered to be resistant to the herbicide glyphosate. Glyphosate also was developed by Monsanto and is marketed under the brand name Roundup. RR soybeans are patented in the United States by Monsanto.[3] Monsanto licenses the RR technology to seed companies, which incorporate it into their conventional soybean varieties and sell the GE seeds to farmers.

Among other things, the patent gives Monsanto and those companies to whom it has licensed the technology more control in setting prices and restricting

[1] U.S. Depart. of Agriculture (USDA), World Agricultural Outlook Board, "World Agricultural Supply and Demand Estimates," June 11, 2003.

[2] For more information on biotech-related issues, see CRS Report RL31970, *U.S. Agricultural Biotechnology in Global Markets: An Introduction;* and CRS Report RS21556, *Agricultural Biotechnology: The U.S.-EU Dispute.*

[3] The patents on RR soybean technology expire on July 10, 2007. U.S. GAO, *BIOTECHNOLOGY: Information on Prices of Genetically Modified Seeds in the United States and Argentina,* GAO/RCED/NSIAD-00-55, January 2000, p.13.

the product's use. For example, U.S. farmers pay a technology fee estimated at $7.44 on each 50-pound bag of RR planting seed.[4] In addition, as part of a signed purchase agreement, U.S. farmers are prohibited from saving seed from harvest for future planting or for resale to other farmers. Despite the additional cost and restrictions, RR soybeans are favored over traditional varieties because they significantly lower production costs, offer more flexibility in crop management, and in many cases increase yields. According to USDA's March 31, 2003, *Planting Intentions Report,* 80% of the soybeans planted in the United States in 2003 will be RR varieties.

The Monsanto company is based in St. Louis, Missouri, but has offices throughout the world where its seeks to market its technology to agricultural producers. Since 1960, a total of 30 countries have approved Monsanto's RR soybean technology for import or planting.[5] However, Monsanto has been unable to obtain patent protection in either Argentina or Brazil. In 1995, Monsanto's application for a patent for RR soybeans in Argentina was rejected. Subsequent applications have not succeeded.[6] In Brazil, the commercial status of GE soybeans (and GE crops in general) remains in dispute in the courts and no patents on RR technology have been issued. The status of GE soybeans in Brazil is particularly important to international oilseed markets since Brazil represents essentially the last potential major source for non-GE soybeans to world markets.

UNIVERSAL ADOPTION OF RR SOYBEANS IN ARGENTINA

Although RR soybean seeds are not patented in Argentina, Monsanto has agreements with other seed firms in Argentina allowing them to use the RR technology in their seeds. As a result, Argentine farmers have access to and have increasingly switched to RR soybean varieties. In 2001, about 90% of Argentina's

---

[4] The estimated technology fee is based on a comparison of GE versus non-GE soybean costs of production presented in the Iowa State University Extension publication, "Estimated Costs of Crop Production in Iowa — 2003," FM 1712, January 2003; and from discussions with experts from Iowa State University and the Iowa Grain Quality Initiative. The fee is usually embedded in the total seed cost and seed costs may vary with marketing margins and sales tactics, making the actual technology fee difficult to assess.

[5] Testimony by Carl Casale, Vice President for Monsanto's North American Agricultural Business; given at Hearing of Senate Foreign Relations Subcommittee on Western Hemisphere, Peace Corps and Narcotics Affairs, May 20, 2003.

[6] Monsanto appealed the initial patent rejection, and an Argentine court overturned the rejection. Monsanto has petitioned for reconsideration of the patent application; however, it appears to remain pending.

soybean crop was reported to be planted to RR varieties.[7] The RR share has grown to nearly 100% of the 2003 crop, according to various news reports.

An Argentine seed law (Act No. 20247; 1973) requires that all seeds must be certified for commercial use.[8] However, Argentine fanners have reportedly ignored this law and routinely save RR soybean seeds for planting or resale. Since the government does not enforce this law, a large black market for RR soybeans in Argentina has developed that keeps seed prices low and discourages any attempts by Monsanto or licensed companies to assess technology fees on RR soybeans. According to Monsanto, it is not feasible to charge a technology fee on soybean seeds in Argentina without patent protection.[9] As a result, Argentine farmers save about $8 to $9 per metric ton on the technology fee.[10] This is a considerable cost advantage over U.S. soybeans in a highly competitive international soybean market.[11]

STATUS OF GE CROPS IN BRAZIL STILL IN FLUX

In September 1998, the Brazilian Biosafety Commission (CTNBio), acting under government authority, approved the commercial planting of RR soybeans. Two major groups opposed to the use of GE crops — Greenpeace and the Brazilian Consumer Defense Institute — immediately filed lawsuits in Brazilian courts challenging the CTNBio approval. In 1999 a lower court issued an injuction suspending the CTNBio approval of commercial planting of RR soybeans before any approved commercial planting actually occurred. The case was appealed to a three-judge panel of the Brasilia Appeals Court, where it has languished.[12] In February 2002, the lead judge of the three-judge Appellate Court

---

[7] Foreign Agricultural Service, USDA. "Argentina Oilseeds and Products Annual," attache report GAIN report #AR1027, May 8, 2001.

[8] GAO(2000), p.5.

[9] Ibid., p. 14.

[10] Assuming a seeding rate of 75 kilograms (or 165.3 pounds) of RR soybean seed per hectare and a technology fee of $7.44 per 50 pounds equals a technology fee of $24.60 per hectare or about $10 per acre. A yield of 2.8 metric tons per hectare (or 41.2 bushel per acre) gives a per metric ton fee saving of about $8.79. The technology fee saving will vary with seed application rates and yields.

[11] During 2002, soybeans traded in a price range of $160 to $215 per metric ton, FOB, Rio Grande, Brazil; $158 to $209 per metric ton, FOB, Buenos Aires, Argentina; and $173 to $225 per metric ton, FOB, Gulf Ports. Source: USDA, Foreign Agricultural Service, "Oilseeds: World Markets and Trade," Table 20, June 2003; and Republica Argentina, Ministry of Agriculture, Livestock, Fisheries, and Food (SAGPyA), [http://www.sagpya.mecon.gov.ar/].

[12] USDA, Foreign Agricultural Service. "Brazil Oilseeds and Products Annual 2003," GAIN report #BR3003, March 10, 2003, p. 5.

publicly announced in favor of commercial planting of RR soybeans. However, a majority of the three-judge panel has yet to render a decision on the commercial use of RR soybeans.

As a result, it remains illegal to plant GE soybeans in Brazil. However, Brazilian farmers are aware of the benefits of RR soybeans and have reportedly smuggled seeds into Brazil from Argentina's black market. Despite the lack of government approval, 70% of the crop in the southernmost state of Rio Grande do Sul is estimated to be planted to RR soybean varieties.[13] USDA estimates that 10% to 20% of Brazil's total soybean crop may be planted to RR. soybean varieties (trade estimates range as high as 30%). As in Argentina, no technology fees are paid by RR soybean growers in Brazil.[14]

More recently, other international events have forced the Brazilian government to temporarily alter its official position on GE soybeans. In June 2001, China—the world's leading importer of soybeans — issued controversial rules governing the use, sale, and importation of GE soybeans. Under the rulings, China will only accept GE soybeans that have been approved for export by the source country, along with certain other conditions. On January 10, 2003, China rejected Brazil's initial application to export GE soybeans, in part because Brazil does not officially recognize the domestic production and export of GE soybeans. In late March 2003, under pressure from producer groups, the Brazilian government announced temporary Regulation 113 (R113) as an interim measure allowing official sales of RR soybeans from the current crop for both domestic uses and export. R113 expires in March 2004, after which Brazilian growers are expected to comply with the current law. In other words, barring any court ruling in the interim, it will be illegal to plant RR soybean varieties in Brazil next season (April 2003-March 2004).

The Lula government has shown no indication of resolving the controversy over commercial legalization of RR soybean in particular, or GE crops in general, anytime soon. News reports indicate that many grower associations within Brazil have already announced their intentions to ignore the existing law and to plant RR

[13] Ibid., p. 38.

[14] The use of pirated RR soybean seeds reportedly has spread to Paraguay, where an estimated 69% of soybean area is planted to RR soybeans *(Food Chemical News,* No. 12, Vol. 45, May 5, 2003, p. 17). Despite pressure from producer groups to legalize RR soybeans, Paraguay's agriculture minister has said that his country would wait and follow Brazil's lead on legalization of commercial GE crops. Paraguay is landlocked and dependent on roads through Brazil to reach international markets. A decision by Brazil against legalizing GE crops could pose serious marketing problems for Paraguay since transportation costs via Argentina are significantly more expensive and their use would likely lower the competitiveness of Paraguay's soybeans in international markets.

soybeans again next season. Brazilian producer groups claim that lack of access to RR technology would place them at a competitive disadvantage in international markets. To date, no significant market premium for non-GE soybeans has emerged in either domestic or international markets sufficient to offset the cost advantages of adopting GE varieties.

Given the rapid growth and widespread use of GE soybeans, legalizing their commercial use may be the only viable solution for Brazil's government. Within Brazil, two principal camps argue against legalizing GE soybeans, but for very different reasons. Some consumer and environmental groups argue that, because the risks associated with GE crops are unknown, they should not be legalized. Peasant groups such as the Landless Workers Movement (MST), on the other hand, are not against GE crops per se, but argue that legalizing GE crops will accelerate large-scale farming and give control over Brazil's agriculture to multinational corporations such as Monsanto.[15] The Lula government remains split on this issue—the Ministry of Agriculture favors prompt legalization while the Minister of Environment is opposed and has asked for an environmental impact study before commercialization is allowed.[16] Both ministers agree that rigid enforcement of existing regulations would mean the incineration of all GE crops and the imprisonment of growers for one to three years at great cost to the country's agricultural sector.

GE Labeling in Brazil

Brazil's food labeling regulation for products containing GE ingredients took effect on December 31, 2001.[17] The law mandates the labeling of all foods for human consumption when more than 4% of the ingredients are derived from GE commodities. R113 has imposed stricter GE labeling conditions on Brazil's food marketing system. Under R113, non-GE soybeans are to be segregated with a 0% tolerance level from GE soybeans.[18] In addition, labeling is required on all shipments into or out of Brazil with a GE soybean presence in excess of 1%.[19]

---

[15] *Inter Press Service,* Rio de Janeiro, Brazil, June 23, 2003.

[16] *Food Chemical News,* No. 3, Vol. 45, March 3, 2003, p. 6.

[17] USDA, Foreign Agricultural Service. "Brazil Oilseeds and Products Annual 2003," GAIN report #BR3003, March 10, 2003, p. 38.

[18] World Perspectives Incorporated, "Soybean Biotechnology Update," April 1, 2003.

[19] *Food Chemical News,* No. 10, Vol. 45, April 21, 2003, p. 23. Argentina's agricultural minister has expressed concern that the labeling requirement under R113 will raise production costs for Argentine producers whose primary market for corn, wheat, and many other agricultural products is Brazil. Argentina's GE soybeans are processed into a wide range of exported food

Since improper labeling is subject to a severe fine, and since Brazil's marketing system is not set up to handle such strict segregation requirements, it is likely that all soybeans passing through Brazil's marketing system will have to be labeled as having GE content.

It is reported that Brazil's National Agriculture Federation (NAF) estimates the cost of testing the current 2002/03 crop for GE content at about $277 million. The NAF says that passing this cost on to consumers would make it impossible for Brazil to remain competitive in the global soybean market.[20]

International Market Implications

According to many market analysts, the decision on the status of GE crops in Brazil will have significant market implications, especially for global soybean markets. First, if Brazil permanently legalizes GE soybeans, nearly all (about 90%) of the world's internationally traded soybeans will be of GE varieties. Second, a decision in favor of GE crops will be nearly irreversible because of mixing in the distribution and transportation systems. Some analysts suggest that the current widespread planting of GE crops in Brazil is already irreversible. In short, a decision in favor of GE soybeans by Brazil could do much to moot the debate about whether or not GE soybeans should be labeled or even traded because there simply would not be any major international supplier of non-GE soybeans left.

U.S. Perspective

According to the U.S. Trade Representative (USTR), the U.S. is committed to a policy of promoting increased intellectual property protection, both through the negotiation of free trade agreements that strengthen existing international laws and through use of U.S. statutory tools as appropriate.[21]

The U.S. Administration's position regarding GE crops is that, not only are food products made from GE crops as safe as their conventional counterparts, but their production has the potential to spur agricultural productivity while

products. In addition, an important share of Argentina's corn crop is planted to GE corn varieties.

[20] *Pesticide & Toxic Chemical News,* No. 25, Vol. 31, April 14, 2003, p. 38.

[21] USTR, 2003 *Special 301 Report,* May 2, 2003, p. 2.

benefitting the environment.[22] Congress has generally supported this position. For example, both the Senate (S.Res. 154) and the House (H.Res. 252) have passed resolutions in support of the Administration's dispute settlement case at the World Trade Organization (WTO) brought against the European Union's ban on imports of GE crops.

In recent hearings by the Senate Foreign Relations Subcommittee on Western Hemisphere, Narcotics, and Peace Corps on May 20, 2003, to discuss opportunities for U.S. agriculture in multilateral and bilateral agricultural trade negotiations in the Western Hemisphere, the IPR issue of RR soybean piracy in Brazil was raised in testimony given by the American Soybean Association and Monsanto.[23] In the absence of patent protection in Argentina or Brazil, accusations of IPR violation may be difficult to sustain in a court of law. However, both Argentina and Brazil are members of the WTO and, as such, have agreed to abide by the WTO agreement on Trade-Related Aspects of Intellectual Property Rights (TRIPS Agreement). As a result, if Monsanto were able to eventually obtain patent protection for the RR technology in either Argentina or Brazil, Monsanto could then seek recourse for IPR infringement via the legal systems of those countries per requirements of the WTO TRIPS agreement.

---

[22] USTR, online resources: biotechnology, June 26, 2003, [http://www.ustr.gov/new/biotech.htm].

[23] Testimony by Bart Ruth, Chairman, American Soybean Association, and Carl Casale, Vice President for Monsanto's North American Agricultural Business; given at Hearing of Senate Foreign Relations Subcommittee on Western Hemisphere, Peace Corps and Narcotics Affairs, May 20, 2003.

Chapter 12

BIOTECHNOLOGY AND AGRICULTURAL TRADE[‡]

Charles E. Hanrahan

ISSUE

Differences between the United States and its trading partners over genetically engineered (GE) crops and food products that contain them pose a potential threat to, and in some instances have already disrupted, U.S. agricultural trade. Underlying the conflicts are pronounced differences between the United States and several important trading partners in consumer attitudes about GE products and their potential health and environmental effects.

BACKGROUND

Widespread fanner adoption of bio-engineered crops in the United States makes consumer acceptance of GE crops and foods at home and abroad critical to producers, processors, and exporters. U.S. farmers use GE crops because they can reduce input costs or make field work more flexible. Supporters of GE crops maintain also that the technology holds promise for enhancing agricultural productivity and improving nutrition in developing countries. U.S. consumers,

---

[‡] Excerpted from CRS Report BTR-2046. Sept. 17, 2001.

with some exceptions, have been generally accepting of the health and safety of GE foods while more attention has been paid to the environmental consequences of planting GE varieties. In contrast, in the EU, Japan, South Korea, and elsewhere, consumers, environmentalists, and some scientists maintain that the long-term effects of GE foods on health and the environment are unknown and not scientifically established. The EU, in particular, insists that precaution should be used in approving and regulating GE foods.

U.S. regulations for GE foods have facilitated their introduction into U.S. agriculture and food processing. The principle has been that GE foods are no different from non-GE foods; therefore, existing regulations for approving foods are appropriate and adequate. Labeling with respect to GE content is not required, except where there is a significant difference between the conventional and the GE food product (for example, the presence of an allergen). The EU, Japan, South Korea, China, Australia, and New Zealand either have or are establishing mandatory labeling requirements for products containing GE ingredients. Japan, the EU, South Korea, and China are respectively the first, fourth, fifth, and sixth largest overseas markets for U.S. agricultural exports. Moreover, the EU maintains a *de facto* moratorium on approvals of new GE crop varieties as it develops further EU-wide legislation for approving and regulating GE products, for tracing GE crops through the marketing chain, and for labeling products that contain GE ingredients.

The U.S. food and agriculture sector faces the challenge of responding to consumer demand, especially overseas, for products differentiated as to their GE or non-GE content. U.S. industry is assessing the costs and benefits of separating GE from non-GE crops and of preserving crop identity in the marketing chain. U.S. regulators are making changes to facilitate voluntary labeling and/or enhance systems for certifying statements about the GE content of foods. U.S. agribusinesses also are seeking to influence EU regulations for tracing GE foods through the marketing chain and for labeling GE foods. Some also have suggested that EU regulations of GE foods could be challenged in WTO dispute settlement.

LEGISLATION

Biotechnology issues seem likely to receive attention in the 107th Congress, particularly in light of the recent discovery of the presence in processed corn products for human consumption of the GE corn variety (StarLink™), which had been approved only for feeding livestock. Two bills relating to genetically engineered foods have been introduced in the current Congress. *H.R. 115* provides for a program of public education about biotechnology in food production, while *H.R. 713* requires the Secretary of Agriculture to conduct, through the National Academy of Sciences, a report on the safety and monitoring of GE foods.

In the 106th Congress, proponents of GE foods introduced bills (*S. 19*, *S. 101*, *S. 566*, and *H.R. 817*) to require U.S. trade negotiators to address "any unjustified restrictions or commercial requirements affecting new technologies, including biotechnologies" in multilateral trade negotiations. Proponents of labeling introduced bills that would have required labeling of foods as to their GE content (*S. 2080. H.R. 3377*) *S. 3184* would have required pre-market consultation and approval of GE foods.

Chapter 13

U.S. AGRICULTURAL BIOTECHNOLOGY IN GLOBAL MARKETS: AN INTRODUCTION[*]

Geoffrey S. Becker and Charles E. Hanrahan

OVERVIEW

U.S. farmers are widely adopting biotechnology, growing genetically engineered (GE) crops — mainly corn, soybean, and cotton varieties — to lower production costs and reduce labor requirements.[1] In 2002, 66% of an estimated 145 million acres planted to GE crops worldwide were in the United States, according to the International Service for the Acquisition of Agri-biotech Applications (ISAAA).[2]

U.S. crops where GE varieties are common are highly dependent upon export markets. According to USDA, approximately 40% of all U.S. soybeans, 20% of all corn, and 45% of all upland cotton production is exported. These crops and

---

[*] Excerpted from CRS Report RL31970. June 19, 2003.

[1] For data and details on U.S. plantings of GE crops, see CRS Report RS21381, *Adoption of Genetically Modified Agricultural Products.*

[2] ISAAA, *Global Status of Commercialized Transgenic Crops: 2002.* ISAAA is a private organization that promotes agricultural biotechnology. Argentina grew 22% of world acreage, following by Canada and China at 6% and 4%, respectively. Twelve other countries also grew some GE crops on a limited number of acres, ISAAA reported. Although no GE crops are approved for commercial use in Brazil, an estimated 20-30% of its 40-42 million acres of soybeans were believed to be GE varieties in 2002, according to various USDA and private analysts.

their major products accounted for more than a fourth ($13.3 billion) of the total annual average annual value of $52.6 billion for all U.S. agricultural exports during calendar years (CY) 2000 to 2002. Not all of these soy, corn, and cotton exports were GE varieties, but the U.S. marketing and regulatory systems do not distinguish between (approved) GE and non-GE varieties. Commingling bulk commodity shipments provides an important cost competitive advantage for U.S. handlers in domestic and world markets, according to industry analysts.

These analysts observe that U.S. adoption of agricultural biotechnology has been facilitated by the current U.S. regulatory system. The basic federal guidance for regulating biotechnology products is the Coordinated Framework for Regulation of Biotechnology (51 *Fed. Reg.* 23302) published in 1986 by the White House Office of Science and Technology Policy (OSTP). One of its key principles is that genetically engineered products should continue to be regulated according to their characteristics and unique features — not according to their method of production. Thus, if a food product produced through biotechnology is determined to be substantially equivalent to one produced by more conventional means, that food is subject to no additional (or no different) regulatory processes. Once approved, food products do not have to be labeled as to whether or not they contain any genetically modified organisms (GMOs), except to the extent a GE food is substantially different (e.g., contains an allergen or has a changed nutritional content). However, marketers are free to make such claims, one way or the other, so long as the labeling is truthful. The framework maintains that new biotechnology products are regulated under existing federal statutory authorities, all of which were conceived and enacted before the advent of commercial agricultural biotechnology.[3]

The problem for U.S. agriculture is that many countries remain wary of agricultural biotechnology, including those in the European Union (EU), where consumer and environmental organizations have been vocal in expressing concerns about the safety of GE crops and animals. The EU and other important U.S. trading partners have adopted widely divergent approaches to regulating biotechnology. As a result, U.S. exporters are encountering barriers to their products in these markets.

For example, since 1998, the EU, the fourth largest foreign market for U.S. agricultural products, has maintained a de facto moratorium on approvals of new GE crop varieties. In May 2003, the United States, Canada, and Argentina began a formal challenge of the EU policy in the World Trade Organization (WTO),

[3] For details see: CRS Report RL30198, *Food Biotechnology in the United States: Science, Regulation, and Issues.*

contending that it violates international trade agreements and also has fueled unwarranted concerns about the safety of agricultural biotechnology throughout the world (see page 7). The EU counters that it must protect its consumers by exercising the so-called precautionary approach, which says that if scientific evidence is insufficient or inconclusive regarding a practice's or product's potential dangers to human or environmental health, it should be more vigorously regulated or even prohibited if there are reasonable grounds for concern, thus providing a safeguard against future unforeseen problems. Under this approach — which is also being emulated somewhat in other important U.S. markets such as Japan and South Korea, for example — the products of biotechnology are deemed to be inherently different than their conventional counterparts.

Even some countries that grow GE crops are imposing their own approval and labeling regulations for GMOs. For example, China is planting its own GE cotton and other crops, but has imposed a temporary GMO import regime while it develops permanent new rules for approval and labeling of GMO farm products. This has created uncertainty about continuing access to this important U.S. export market.

Some U.S. producers, who otherwise generally have supported biotechnology, have expressed trepidation about expansion to more GE varieties due to such foreign market uncertainties. Their concerns have been evident as biotechnology companies work toward government approvals of GE wheat. Although some growers are eager to plant the new varieties when they become available, others do not want commercialization approvals until there is wider global acceptance.[4] Wheat growers, too, depend heavily on exports; USDA reports that 56% of production is exported. The average annual value of U.S. wheat exports was $3.6 billion (7% of all agricultural exports) in CY2000-2002.

[4] GAO; Pew Initiative on Food and Biotechnology. *USDA May Set Strict Rules for Biotech Wheat,* March 17, 2003. Also, "USDA Seeks to Avoid Disruption in Wheat Exports After First GMO Approval," *Inside U.S. Trade,* March 21, 2003.

International Oversight of Agricultural Biotechnology

National Regulation[5]

Around the world, countries have taken widely divergent approaches to regulating the products of agricultural biotechnology. Critics have noted that many, particularly in the developing world, have adopted no coherent policy at all.[6] Among those that do, each has its own system. Although multilateral attempts to harmonize GMO regulation have been under way for several years, internationally recognized standards still are evolving.

The disparity of national regulatory responses has been characterized as a "renationalization" of agri-food safety regulation, and "a sharp break from the international food safety system that evolved over the past 100 years, where importers tended to accept the food and environmental safety judgments of regulators from those countries developing and exporting the products."[7] Others argue that scientific knowledge about biotechnology's effects on food and environmental safety is still incomplete. In the meantime, they argue, countries have a right and obligation to take the "precautionary approach" to regulation. Various countries' national biotechnology regulation is occurring within two broad categories:

Approval to Commercialize New Products of Agricultural Biotechnology

The United States, Canada, Japan, Mexico, Argentina, and South Africa are among the countries where developers seeking approvals of new GE products have found a relatively flexible regulatory environment. To this list might be added India and China, where considerable research has been conducted on agricultural biotechnology and domestic approvals appear to be increasing, but where, nonetheless, imports of GE crops may face obstacles.[8] In Australia, the

---

[5] Sources for this section include: International Food Policy Research Institute (IFPRI). *Policy, National Regulation, and International Standards for GM Foods,* by Peter W.B. Phillips, professor of agricultural economics, University of Saskatchewan, January 2003 (hereinafter cited as IFPRI/Phillips); World Health Organization, *Twenty Questions on Genetically Modified (GM) Foods,* 2002; and General Accounting Office (GAO), *International Trade: Concerns Over Biotechnology Challenge U.S. Agricultural Exports* (GAO-01-727), June 2001.

[6] See for example the testimony of John Kilama, President, Global Bioscience Development Institute, before the House Committee on Agriculture, March 26, 2003.

[7] IFPRI/Phillips.

[8] Reports by USDA's Foreign Agricultural Service (FAS) and by other trade officials suggest that India's seeming reluctance to accept imports of GE soybean oil, and China's imposition of

EU, and New Zealand, approvals have slowed. A total of nearly 30 countries reportedly have developed relatively coherent (although in the case of some countries, either somewhat ambivalent or not necessarily fully transparent) policies with regard to GE approvals.

Systems for Labeling and Traceability

In Argentina, Canada, Hong Kong, and the United States, labeling products as to whether or not they are from GE-derived crops generally is voluntary. However, the EU and more than 20 countries have either adopted, or announced plans for, mandatory labeling of GMO products. A number of them, for example, Australia, Brazil, China, Indonesia, Japan, New Zealand, Saudi Arabia, South Korea, and the United Kingdom (an EU member), have formally implemented their mandatory labeling regimes. Many others, however, have not fully outlined their proposals or indicated when they might take effect; some also apparently lack implementing structures.[9] Generally, mandatory labeling also necessitates accompanying traceability requirements, meaning that the marketing system must have documentable ability to trace the presence or absence of GMOs through each step from farm to point of sale.

Multinational Oversight

The wide range of national approaches to GMO regulation is in part due to the fact that an international consensus on how to regulate agricultural biotechnology is still evolving. Agricultural biotechnology and trade policy analysts throughout the world generally agree that an international consensus on how best to regulate agricultural biotechnology is not only desirable but necessary. Attempts have been under way for some years in a variety of multinational institutions to develop more harmonized standards for ensuring the safety of GE crops and foods, and to establish ground rules for testing, enforcement, and dispute resolution. It is not yet clear which of these international bodies will have the greatest influence in resolving these questions, although it appears likely that most, if not all, will have some role.

---

additional testing requirements for imported GE crop varieties, may have more to do with trade protectionist tendencies than with food and environmental safety concerns.

[9] IFPRI/Phillips. Also see: Sheldon, Ian, Ohio State University, and Tim Josling, Stanford University, *Biotechnology Regulations and the WTO,* International Agricultural Trade Research Consortium Working Paper 02-2, January 2002.

The IFPRI/Phillips paper (see footnote 5) observed that "Despite the substantial effort being undertaken, there is no common view on the goal of international regulation. While most agree that safety is the bottom line, few can agree on what that means, whose opinions should hold the most weight (scientists' or citizens'), or how to handle nonsafety issues such as social, economic, or ethical concerns."

A wide array of multinational bodies has, or claims, a role in some facet of biotechnology regulation and standard-setting (see **Table** 1). Several are longstanding and largely science-based organizations, including the U.N. World Health Organization (WHO), the U.N. Food and Agriculture Organization (FAO), the food standards-focused Codex Alimentarius Commission, the International Plant Protection Convention (IPPC), and the International Epizootics Organization (OIE). The WTO and the Organization for Economic Cooperation and Development (OECD) are primarily (although not necessarily exclusively) concerned about free trade.

Codex, for example, was created in 1963 by FAO and WHO to develop internationally recognized, science-based food standards and guidelines and related materials. Its main objectives are to protect consumer health, ensure fair trade, and promote coordination of food standards work undertaken by governmental and non-governmental organizations. Codex standards are used to evaluate national food regulations and to settle sanitary and phytosanitary (SPS) related trade disputes in the WTO.

In 1999, the Codex established the ad hoc Intergovernmental Task Force on Foods Derived from Biotechnology. The task force was charged with considering health and nutritional implications, and with developing standards, guidelines or recommendations, as appropriate, for such foods. The task force held its fourth and final meeting in March 2003, when it completed work on the last of three framework documents, *Draft Principles for the Risk Analysis of Foods Derived from Recombinant-DNA Microorganisms.* The task force's documents were to be presented for approval by the full Commission at its July 2003 meeting.

U.S. officials expressed satisfaction with the outcome of the final session, although deliberations over the guidelines often were contentious. At the last meeting, for example, an open discussion was held on rules for traceability of GE products. The EU led by France was advocating that they encompass rules for full traceability of GE products. The United States was arguing that traceability is acceptable only for food safety/public health purposes, not merely to preserve the identity of specific products, including those resulting from biotechnology. The

U.S. view is that these types of questions should be considered in other Codex committees because they are not unique to biotechnology.[10]

The 2000 Cartagena Biosafety Protocol is a newer, environmentally-focused effort devoted to the safe transfer, handling and use of bio-engineered products crossing international borders. The United States is not a party to the Protocol but is participating to ensure that its provisions do not undermine such trade in GM-derived products.

Table 1. Multinational Institutions Involved in Biotechnology

Institution (No. of members)	Purpose/Activity	Lead U.S. Agency
(Cartagena) Biosafety Protocol (50)	January 2000 agreement, under the U.N. Convention on Biological Diversity, on movement and use of living modified organisms. 50 countries, the required number for entry into force, have ratified the treaty. U.S. is not a party to the Convention or Protocol, but is working with ratifiers, and others, to ensure implementation does not harm U.S. exports.	State
Codex Alimentarius (165)	Sets science-based international food safety standards recognized by WTO [in the Sanitary & Phytosanitary (SPS) Agreement]. A Codex task force has devised guidelines for assessing risks; final Codex approval is anticipated.	FDA
Food & Agriculture Organization of U.N. (183)	Interested in food security benefits of biotechnology vs. potential safety risks. FAO Commission on Genetic Resources for Food & Agriculture advises on availability and use of genetic resources for food and agriculture; equitable sharing of benefits — goals of the FAO Treaty on Plant Genetic Resources for Food and Agriculture. The U.S. signed it in November 2002; wants to ensure U.S. agriculture and biotechnology are not adversely affected.	USDA
International Plant Protection Convention (120)	Prevents spread of plant pests & pathogens; its standards are recognized by the SPS Agreement.	APHIS
International Epizootics Organization (162)	Prevents spread of animal pests & pathogens; its standards are recognized by the SPS Agreement.	APHIS
Organization for Economic Cooperation & Development (30)	Composed of developed democracies; fosters market economies, free trade. Promotes international harmonization of biotechnology regulations through development of consensus documents and information, and through outreach.	Depends on issue
World Health Organization (192)	Has main responsibility for health, safety aspects of Codex, including biotechnology. Commissioned an "evidence based" study of human health aspects of GMOs; comments on draft were due in March 2003.	HHS
World Trade Organization (145)	Oversees world trade rules now governed by the 1994 Uruguay Round (UR) Agreements, including trade dispute settlement. The UR predates most agricultural biotechnology, but the U.S. believes such issues can be resolved through the UR SPS and TBT Agreements. WTO also is the forum for the current trade negotiations under the Doha Development Agenda, where agriculture, including, potentially, biotechnology, is on the table.	USTR
(Bilateral Efforts)	The U.S. increasingly works with individual countries to harmonize regulations, educate policymakers and consumers, and eliminate barriers to trade related to biotechnology (see text).	Depends on issue

Sources: IFPRI; USDA.

Agricultural biotechnology was not a major element of international trade during the last comprehensive round of multilateral trade negotiations and therefore is not directly addressed in the 1994 Uruguay Round (UR) agreements. However, the U.S. position has long been that the UR Agreement on Agriculture, the Agreement on the Application of Sanitary and Phytosanitary (SPS) Measures,

[10] For details see: USDA, Food Safety and Inspection Service. Report of the United States Delegate, Fourth Session, Codex Ad Hoc Intergovernmental Task Force on Foods Derived from Biotechnology, Yokohama, Japan, March 11-14, 2003, which can be viewed at: http://www.fsis.usda.gov/OA/codex/rep_bt03.htm.

and the Technical Barriers to Trade (TBT) Agreement provide adequate guidance for dealing with trade in GMOs. Taken together, these agreements seek to reduce or eliminate all unnecessary or arbitrary barriers to trade, and ensure that SPS or technical measures (e.g., GMO labeling) are transparent, scientifically defensible, and based on accepted international standards.[11] The U.S. proposal in the current Doha round of negotiations in the WTO is silent on biotechnology itself but presumes a continuation of this position.

SELECTED U.S. CONCERNS

U.S. officials say they are actively engaged multilaterally and bilaterally to ensure that any national or international standards are consistent, transparent, based on scientific principles, and compliant with international trade rules (e.g., those administered through the WTO). However, they face difficulties in reconciling the current U.S. approach with the opposing perspectives of other influential countries like those in the EU and elsewhere. Following are discussions of selected issues in the international arena which have challenged U.S. supporters of agricultural biotechnology.

The EU's De Facto Moratorium on GMO Approvals[12]

On May 13, 2003, the U.S. Trade Representative (USTR) and the U.S. Secretary of Agriculture announced that the United States, Canada, Argentina, and Egypt were filing a case before the World Trade Organization against the EU's de facto 5-year moratorium on approving new agricultural biotechnology products. (Egypt subsequently decided against filing with the United States.) The Administration was under increasing pressure from key Members of Congress and farm groups to launch such a case. U.S. agricultural interests contend that not only have these policies blocked their exports to the EU, but also fueled unwarranted concerns about the safety of agricultural biotechnology throughout the world.

---

[11] See also ERS, "Biotechnology in Agriculture Confronts Agreements in the WTO," in *Agriculture in the WTO* (WRS-98-4), December 1998.

[12] Sources include: USTR, *2003 National Trade Estimates Report on Foreign Trade Barriers;* Pew Initiative on Food and Biotechnology, *U.S. vs. EU: An Examination of the Trade Issues Surrounding Genetically Modified Food,* 2002; European Commission, *Questions and Answers on the regulation of GMOs in the EU,* March 4, 2003; and Sheldon, Ian, and Tim Josling, *Biotechnology Regulations and the WTO,* International Agricultural Trade Research Consortium, Working Paper #02-2, January 2002.

EU officials say they have been moving as quickly as possible to reinstate biotechnology approvals while trying to reassure their consumers regarding safety issues. There is concern among some policymakers that the filing of a WTO case will only escalate trade tensions between the United States and the EU, where the value of two-way agricultural trade amounts to more than $14 billion annually.

Background

With minor exceptions, the EU has approved no agricultural biotechnology products since 1998, even though it has an elaborate approval process in place, and, in October 2002, implemented revisions to that process aimed at reassuring its member states and the public about the safety of its regulatory system. Approximately 13 products in early 2003 were awaiting approval, some for as long as 6 years. A majority of EU states has effectively blocked the release of any new GE crops into the environment. These states — France, Belgium, Luxembourg, the Netherlands, Germany, Austria, Italy, Ireland, Greece, Spain, Portugal, and Finland — say they will not implement the EU-wide legislation for approvals until new, stricter, regulations for labeling and tracing GM-containing products also are adopted. Although the European Commission first approved these separate proposals in July 2001, the lengthy EU decision-making and implementation process make it unlikely that these proposals will come into force any earlier than some time in 2004.

In the 3 years before the de facto ban, U.S. corn exports to the EU averaged approximately $300 million annually (Spain and Portugal were the largest EU importers). Since then, they have declined to less than one-tenth of that value annually—the result, according to analysts, of the EU's moratorium on the approval of new corn varieties already approved in the United States. Although one variety of biotech corn was approved by the EU prior to the moratorium, other approved varieties are being grown in the United States, making exports of any U.S. corn (which is not separated as to GM or non-GM varieties) to the EU impractical. U.S. soybean exports to the EU have declined, too, from an average value of approximately $2.2 billion in the 3 years before the ban to about $1.2 billion annually (2000-2002), according to USDA data. Unlike biotech corn, only one variety of biotech soybeans has been widely grown in the United States; the EU had approved this GE crop under its regulatory process prior to the moratorium. (The EU also had approved a number of other GMOs prior to the 1998 moratorium.)

The WTO Case

The U.S.-led case began with a request for 60 days of formal consultations with the EU. In addition to the three main filers, Australia, Chile, Colombia, El Salvador, Honduras, Mexico, New Zealand, Peru, and Uruguay expressed their support for the case. On June 19, 2003, a USTR spokesman issued a statement announcing that the consultations were not successful, and that the WTO will now be asked to establish a dispute settlement panel, the next step. At any rate, resolution of the dispute could take up to 18 months.

The United States and its allies argue in part that the EU moratorium violates the WTO Agreement on the Application of Sanitary and Phytosanitary (SPS) Measures. The SPS agreement permits countries to regulate crops and food products to protect health and the environment, but their rules must be scientifically justified and approval procedures must be operated without undue delay. U.S. interests contend that there is no scientific evidence that GM-derived food and feed crops are substantially different from, or any less safe than, conventional varieties, a conclusion they say even European scientific authorities have reached. The moratorium has not only impacted U.S. exports to the EU, but also caused other countries—particularly in the developing world — to shun biotechnology, which, the United States has argued, holds great promise for vastly improving agricultural productivity and feeding growing populations (see also page 11).

EU officials counter that they have shown good faith in moving as quickly as possible to restart the approval process. In early April 2003, the EC threatened to take legal action at the European Court of Justice against the 12 member states if they fail to implement the GE approval legislation once the new labeling and tracing rules are approved. Nonetheless, EU officials have told the United States that their cautious approach to regulating agricultural biotechnology is necessary to restore confidence among European consumers, who maybe more wary of changes in how their food is produced due to a series of recent food safety crises.[13]

It is unclear how a WTO dispute panel might rule on the U.S.-EU case, in part because agricultural GMOs are not explicitly recognized in the Uruguay Round trade agreements, which were concluded before the advent of widespread

[13] For example, during the 1990s, bovine spongiform encephalopathy (BSE, or "mad cow disease") emerged in the United Kingdom and spread to other parts of Europe. In 1999, high levels of dioxin were found in meat products and eggs originating in Belgium. Pew concluded: "Although these crises have not been caused by GE food, GE food has been caught up in the general suspicion about food safety." See: *U.S. vs. EU: An Examination of the Trade Issues Surrounding Genetically Modified Food.*

agricultural biotechnology. If the United States wins, it would validate the basic principles of the SPS agreement and could discourage other countries from emulating the EU regulations. The United States has pointed out that many EU farmers themselves would like to be planting and selling GE crops. However, a win might not open EU markets to more GE imports; the United States might simply have to settle for some form of alternate compensation. Some have suggested that a win could create a backlash among the European public and governments who view the United States as forcing biotechnology on unwilling consumers.

Labeling and Traceability Concerns

As noted, a number of foreign markets for U.S. farm products have been developing systems that will require most or all GE-derived products to be labeled as to their GE content. In the U.S.-EU biotechnology dispute, the U.S.-led WTO challenge to the EU moratorium on GM approvals does not involve the emerging EU labeling regulations. These regulations, if approved as proposed, would require all food, feed and processed products from GMOs to be labeled, even if they no longer contain detectable traces. A tolerance level for non-GMO foods, feeds, and processed products of 0.9% is set for allowable "adventitious presence" (AP)—that is, unintended, low-level presence—of an EU-approved GE substance. All products with more than 0.9% must be labeled as GM. The allowable level for unapproved GE varieties that received a positive EU risk assessment during the moratorium is 0.5% for 3 years, after which it drops to 0%. (The current EU regime requires labeling if ingredients contain more than 1% GMO protein or DNA.)

U.S. officials so far have challenged formally only the EU moratorium, but a similar effort against the labeling and traceability rules is possible in the future. U.S. agricultural interests argue that, even if the GMO approval moratorium is lifted, the new labeling and traceability rules are themselves unworkable and unnecessary, and will continue to discriminate against U.S. exports. Compliance with the EU labeling rule would require segregation of GE crops and foods derived from them from the time they are planted all the way through the processing and marketing chain. This would entail prevention of pollen drift from GE to non-GE fields; and difficult and costly handling procedures such as using separate equipment, storage, and shipping containers, or at least painstakingly cleaning them. U.S. interests argue that food companies forced to label accurately all GE products could face huge risks and liabilities. All of these problems would discriminate against U.S. shipments, they add. Material from Virginia Cooperative Extension asserts:

> Segregating crops at every step during growth, storage, transportation, processing, and exporting would be very difficult and costly to implement. Dr. Susan Harlander, a former vice president of Pillsbury Company, illustrated the situation with the following example: a medium-sized food company can have more than 6,000 products that contain 8,000 ingredients from 1,000 suppliers that move through 30 processing plants on their way to being exported to as many as 100 countries. Implementing a system to track all of these ingredients from their source (as far back as the farm) to the final destination is a daunting task that would cost billions of dollars and even then it may not be infallible.[14]

According to Virginia Extension, among the issues to be resolved on labeling and detection of GE material in crops and foods are: the scientific methods used to screen the product; the sensitivity of the detection system; what would actually be detected (DNA or protein); what agency would manage and regulate the testing; and who would be held accountable if a product were mislabeled.

The Biosafety Protocol[15]

The Cartagena Biosafety Protocol, an outgrowth of the 1992 Convention on Biological Diversity (CBD), was adopted in January 2000 and will take effect this year, now that the required 50 nations have ratified it. The United States is not a party to the 1992 CBD, and therefore cannot be a party to the Protocol. However, it has actively participated in the negotiations over the Protocol text and in countries' preparations for implementation.

The Protocol addresses a major area of concern not resolved by the parent CBD in 1992—the safe handling, transfer and transboundary movement of bio-engineered organisms and products. It enables countries to obtain information about biotech organisms (i.e., "living modified organisms" or LMOs) before they are imported into those countries. It recognizes each country's right to regulate such organisms so long as they are consistent with existing international

---

[14] Vines, Randy, Virginia Cooperative Extension Specialist, Biotechnology Information. *The Regulation of Biotechnology*. Publication 443-006, May 2002. To view the material, see [http://www.ext.vt.edu/pubs/biotech/443-006/443-006.html]. See also GAO, *International Trade: Concerns Over Biotechnology Challenge U.S. Agricultural Exports*.

[15] Sources: CRS Report RL30954, *Biosafety Protocol for Genetically Modified Organisms: Overview;* U.S. State Department, *Fact Sheet: Update on the Cartagena Protocol on Biosafety*, July 11, 2002; and USDA's notes accompanying its FY2004 budget request.

agreements, and provides for an international clearinghouse for the exchange of LMO information.

The Protocol establishes the use of "Advance Informed Agreements" (AIA) between the importing and exporting parties that cover the first transboundary movement of any GMO. The purpose of AIAs is to ensure that recipient countries have the opportunity to assess environmental risks associated with the importation of biotechnology products. The Biosafety Protocol creates the procedure by requiring exporters to seek consent from importers before the first shipment of a GMO is introduced into the environment (applies to seeds for planting, fish for field release, and microorganisms for environmental bioremediation). In addition, article 11 also requires that bulk shipments of GMO commodities for food, feed, or processing be accompanied by declarations stating that such shipments "may contain" GMOs and are "not intended for intentional introduction into the environment."

One section of the Protocol states that "Lack of scientific certainty due to insufficient relevant scientific information and knowledge regarding the extent of the potential adverse effects of a living modified organism on the conservation and sustainable use of biological diversity in the Party of import, taking also into account risks to human health, shall not prevent that Party from taking a decision, as appropriate, with regard to the import of that LMO ... in order to avoid or minimize such potential adverse effects." Defenders of the application of this approach (essentially the "precautionary principle") maintain that it is only "a temporary mechanism" that gives time for scientific inquiry. Policymakers in Europe recognize, as do those in the United States, the need for an assessment of risks based on accepted scientific facts. However, critics worry that adopting the precautionary principle in formal agreements like the Protocol may create false public expectations of absolute safety and the demand for zero environmental risks. Further, some critics maintain, the precautionary approach can be used as disguised protectionism — as they say already has occurred on the part of the EU in blocking U.S. beef imports containing growth hormones and in delaying approvals of new GMOs.

Implementation of the AIAs, and of provisions on documentation that countries can require for GMOs and GM-containing shipments, also are of major concern to the United States and other GMO exporters. Government and industry officials from the exporting countries have been meeting with each other, and with the ratifying countries, to ensure that bulk grain shipments, processed foods, and other trade in products that may contain GMOs are not disrupted due to confusing, repetitive, or inconsistent new requirements, shipping delays, burdensome paperwork, and the like.

Food Aid and GMOs

Six countries in southern Africa—Lesotho, Malawi, Mozambique, Swaziland, Zambia, and Zimbabwe — are experiencing continuing severe food shortages. In 2002, the United Nations World Food Program (WFP) announced an appeal for food aid to meet the needs of some 14 million people in the six countries. However, provision of food aid to meet the needs of food-short people in the region has been made more difficult and costly by a debate over the presence of genetically modified corn in U.S. food aid shipments. Some of the countries expressed reluctance to accept GE corn on account of perceived environmental and commercial risks associated with potential introduction of GE seeds into southern Africa agriculture. One, Zambia, refused all shipments of food aid with GE corn out of health concerns as well.

A concern of the southern Africa countries is that corn contained in U.S. food aid shipments could be planted either accidentally or intentionally by African farmers and damage future trade with the European Union. Introduction of genetically modified corn into their food systems could mean that crops grown for export or used in livestock feed would contain genetically modified components. As noted, the EU has approved and permits imports of genetically modified soybeans, but has not yet approved imports of most varieties of genetically modified corn. The EU is expected soon to put in place a system of tracing and labeling of food products as to their GE content While livestock products will be exempted from such labeling, some of the African countries express fears that EU consumers already antipathetic to GE foods will reject meat products that might have been produced from animals fed GE grain.

The governments of Zimbabwe and Mozambique have accepted U.S. food aid shipments of corn only if it is milled prior to distribution. Malawi has requested that corn be milled, but according to the U.S. Agency for International Development (USAID), is allowing distribution of whole grain com due to limited milling capacity. Swaziland and Lesotho have expressed reservations about food aid shipments of GE corn, but are accepting them. Only Zambia has rejected any food aid containing bio-engineered gram.

The United States has blamed EU policies for southern Africa countries' views on food aid containing GE products. President Bush, for example, has stated that EU governments, because of their policies on GE products, are hindering the cause of ending hunger in Africa.[16] The United States maintains that genetically modified crops are safe to eat; and that there is little likelihood of GE

[16] Commencement Address to the U.S. Coast Guard Academy, May 21, 2003.

corn entering the food supply of African countries for several reasons, including the fact that bio-engineered varieties of corn are not well adapted to African growing conditions.[17]

Biotechnology and Developing Countries

Relatedly, there is considerable debate over the potential contribution of biotechnology to food security and agricultural development in developing countries. Critics of biotechnology argue that the benefits of biotechnology in developing countries have not been established and that the technology poses unacceptable risks or problems for developing country agriculture.[18] They argue that bio-engineered crops could adversely affect the health of consumers and that genetically engineered crops pose unacceptable environmental risks for agriculture in developing countries. Critics suggest that intellectual property rights (IPR) protection impedes development and dissemination of GE crops in developing countries and also gives multinational companies control over developing country farmers. Proponents, however, say that the development of GE technology appears to hold great promise, with the potential to complement other, more traditional research methods, as the new driving force for sustained agricultural productivity in the 21st century. They maintain that GE foods are safe to consume, that environmental risks are both negligible and manageable, and that IPR difficulties have been exaggerated.[19]

CONGRESSIONAL RESPONSE

Members of the House and Senate Agriculture, House Ways and Means, House Science, and Senate Finance Committees are among those in the 108th

---

[17] For more details, see [http://www.usaid.gov/about/africafoodcrisis/bio_answers.html].

[18] Typical of the arguments against introducing bioengineered crops into developing country agriculture is a report by the Kasisi Agricultural Training Centre (Zambia) and the Jesuit Centre for Theological Reflection entitled *What* is *the Impact of GMOs on Sustainable Agriculture in Zambia,* August 15, 2002, available at [http://www.jctr.orlg.zm]. For arguments in support of introducing bioengineered crops, see: House Committee on Agriculture hearing, *Review of Artificial Barriers to U.S. Agricultural Trade and Foreign Food Assistance,* March 26, 2003, at [http://agriculture.house.gov/hearings/l 081.pdf].

[19] IFPRI, *Biotechnology and Genetic Resource Policies,* Briefs 1-6, January 2003, Brief 3. See also CRS Report RL31568, *Plants, Patents, and Seed Innovation in the Agricultural Industry;* and CRS Report RL31132, *Multinational Patent Acquisition and Enforcement: Public Policy Challenges and Opportunities for Innovative Firms.*

Congress who are closely monitoring biotechnology developments in the international arena, as well as the Administration's response. If this Congress follows the lead of its predecessors, it is more likely to be supportive of agricultural biotechnology in international trade than to seek constraints on GE commodities traded internationally and domestically. For example, the Senate on May 23, 2003, passed, by unanimous consent, a resolution (S.Res. 154) supporting the U.S. action against the EU; a similar House measure (H.Res. 252) was passed June 10, 2003, by a suspension vote of 339-80. Separately, H.R. 2447, introduced June 12, 2003, would establish a federal interagency task force, composed of senior Administration officials, to promote the benefits, safety, and uses of agricultural biotechnology.

The House Agriculture Committee held a hearing on March 26, 2003, where the theme was the EU's moratorium on GMO approvals and its possible influence on decisions by developing African countries to reject or delay U.S. food aid.[20] The House Science Subcommittee on Research held a hearing on June 12, 2003, to examine plant biotechnology research and development challenges and opportunities in Africa.

Earlier, the 107th Congress included biotechnology provisions in the 2002 farm bill (P.L. 107-171) including: a biotechnology and agricultural trade program, aimed at barriers to the export of U.S. products produced through biotechnology (Section 3204); competitive grants for biotechnology risk assessment research (Section 7210); agricultural biotechnology research and development for developing countries (Section 7505); and a program of public education on the use of biotechnology in producing food for human consumption (Section 10802). In addition, legislation that gives the President Trade Promotion or "fast track" authority (P.L. 107-210) contains language providing that U.S. trade negotiators should seek to negotiate rules and dispute settlement procedures that will eliminate unjustified restrictions and requirements, including labeling, of biotechnology products.

Bills also were introduced in the 107th Congress that took other approaches to regulating agricultural biotechnology, and some could reappear in the 108th. They included H.R. 4814, which called for mandatory labeling of GE foods; and other bills in the 107th (H.R. 4812, H.R. 4813, and H.R. 4816) which dealt respectively with legal issues raised by cross-pollination with GE plants, a study of the safety of GE foods, and liability for injury caused by GE organisms. No comparable bills had been introduced as of early June 2003.

---

[20] The transcript *of Review of Artificial Barriers to U.S. Agricultural Trade and Foreign Food Assistance* is at [http://agriculture.house.gov/hearings/1081.pdf].

APPENDIX: U.S. AGENCY ROLES IN AGRICULTURAL BIOTECHNOLOGY TRADE

General[21]

Responsibility for biotechnology trade issues and activities is shared by numerous federal agencies and departments. Officials who deal with them on a day-to-day basis say they communicate often, either individually or in working groups, to share information, resolve jurisdictional questions, and coordinate efforts.

As more difficult or controversial trade-related issues arise, officials from relevant agencies may meet at successively higher levels to discuss how to formulate and implement a policy response. According to U.S. officials, years of experience — along with established statutory authorities and the 1986 Coordinated Framework — have provided them with a strong, if somewhat tacit, understanding on how they will handle any given problem, including which agency should take the lead role. Such experience has fostered an increasingly coordinated, if not yet widely-enunciated, formal approach to biotechnology trade policy, several U.S. officials argued.

Nonetheless, some agencies are more prominent than others. The Office of the U.S. Trade Representative (USTR) is generally regarded as the lead agency on biotechnology trade issues as part of its overall mandate to coordinate U.S. trade policy, including negotiating trade agreements and dealing with major disputes. USTR leads the U.S. delegation of agency representatives at quarterly SPS and TBT meetings convened by the WTO, where biotechnology matters generally are covered. Back home, biotechnology questions pass through the USTR staff-level offices that handle agricultural, SPS, and TBT matters. USTR convenes and coordinates an interagency SPS Trade Policy Staff Committee consisting of representatives from agencies throughout government. This committee meets frequently, but on an "as-needed" basis, on biotechnology trade issues, some of which may arise from ongoing bilateral or multilateral negotiations, and others from disputes with trading partners. USTR also convenes a higher-level SPS Trade Policy Review Group, composed of more senior agency officials, to discuss

---

21 Sources for this section include: telephone interviews with White House, USTR, and USDA officials in late March and early April 2003; and GAO's *International Trade: Concerns Over Biotechnology Challenge U.S. Agricultural Exports*.

and resolve more controversial SPS questions, including those on biotechnology.[22]

However, other agencies take the lead on specific biotechnology issues. The State Department, for example, is at the forefront in dealing with countries implementing the Cartagena Biosafety Protocol. PDA, supported by USDA's Food Safety and Inspection Service (FSIS), leads on Codex matters (see **Table** 1).

On occasion, particularly thorny issues and/or jurisdictional disputes must be resolved at the White House level, which ultimately decides overall U.S. biotechnology policy with input from its Office of Science and Technology Policy (OSTP). The President's Special Assistant for Agricultural Trade and Food Aid, and a biotechnology working group composed of high-level officials under the aegis of the White House National Economic Council (NEC), periodically may be called upon to discuss or referee such issues. The National Security Council also has been involved in discussions, e.g., those concerning whether to challenge formally the EU over its de facto moratorium on new GMO approvals.

Some critics assert that the U.S. response to biotechnology remains largely reactive, ad hoc, and not well coordinated. Thus, on any issue, jurisdictional discussions may precede substantive action, they contend, arguing that a more comprehensive long-term strategy is needed. The Biotechnology Industry Organization (BIO) has reported on its website that USTR has circulated among interested agencies a draft paper spelling out such a strategy. However, as of early June 2003, the paper had not emerged publicly.[23]

USDA

Biotechnology issues and activities, including international ones, are addressed both at the departmental level (across agencies) and within the various agencies themselves. USDA's Biotechnology Coordinating Committee is a staff-level interagency group that has met periodically to share information and work on biotechnology issues of mutual concern (both domestic regulatory as well as trade-related matters). A higher-level Biotechnology Policy Group is convened at least several times monthly by the Secretary of Agriculture's special counsel on trade to discuss policy, receive updates, and map plans for upcoming interdepartmental and international meetings. Secretary of Agriculture Ann

[22] GE labeling concerns are within the purview of the separate TBT Trade Policy Staff Committee and Review Groups, also coordinated by USTR.

[23] Based in part on information from BIO, *including Agriculture Biotechnology International Trade Fact Sheet,* available at [http://www.bio.org/foodag/priorities/tradefacts.asp].

Veneman, on April 8, 2003, named 18 people to a new Advisory Committee on Biotechnology and 21st Century Agriculture (authority for a previous advisory committee expired in 2001). Eleven of the members represent agribusiness companies or agricultural producer groups; the rest are from academia or consumer advocacy organizations.

Section 3204 of the 2002 farm bill (P.L. 107-171) created a new Biotechnology and Trade Program to provide grants for public and private sector projects that will address nontariff barriers to U.S. agricultural exports involving biotechnology, food safety, disease, or other SPS concerns; or that will develop, through bilateral negotiations, protocols on animal health, grain quality, and GMOs. The measure authorizes appropriations of up to $6 million annually through FY2007.

USDA's FY2004 budget proposal recommends new funding of $6.6 million and 20 new staff positions "to support a number of important cross-cutting, trade-related and biotechnology activities of the Department and to ensure that adequate funds are available to address them effectively." Of this amount, $2.1 million would be specifically for work on GMO trade issues, presumably approximately a third of the authorized funding under Section 3204. Of this amount, $600,000 would be for the Foreign Agricultural Service's new biotechnology unit (see below), and the other $1.5 million for U.S. activities in other countries, according to a Department official.[24] (Congress had not yet acted on the FY2004 USDA appropriation by mid-June 2003.) The budget documents note that USDA "faces a growing array of challenges related to biotechnology, including a rapidly expanding number of regulatory, market access, and trade barrier issues." The funds would be appropriated directly to the Secretary to provide her with flexibility in allocating them among various agencies.

At the agency level, the **Foreign Agricultural Service** (FAS) facilitates and promotes agricultural trade, including trade in the products of biotechnology. FAS activities include education, training, technical assistance, and issue resolution. In recent years, FAS attention has been shifting increasingly from biotechnology capacity building (e.g., technical assistance on the science and application of biotechnology in agriculture), to monitoring and dealing with foreign regulatory policies and restrictions on trade in GMO products. FAS programs include:[25]

---

[24] USDA also has asked Congress for $500,000 in FY2004 to conduct a series of regionally-based short courses on the specifics of the Cartagena Protocol to ensure that countries implement it in a "science-based, transparent, and non-discriminatory" manner.

[25] See the FAO website, *Biotechnology and U.S. Agricultural Trade, Training and Capacity Building,* at [http://www.fas.usda.gov/itp/biotech/development.html].

- An Overseas Biotech Training/Education Program involving seminars, symposia, and educational materials on various biotechnology issues aimed at foreign educators, public officials, and others;
- Cochran Fellowship Program Biotechnology Training, offering short-term U.S. training for foreign scientists, regulators, journalists, and policymakers, to provide information about the technology's benefits and about U.S. regulation;
- Biotechnology Short Courses, a new series of quarterly, two-week programs in the United States for foreign participants intended primarily to impart "biotechnology's relationship to market access and trade in agricultural products, and the factors influencing that relationship";
- Capacity Building for the Seed Trade Industry to expand seed trade with Eastern Europe, Africa, and Asia;
- Biotechnology Research Capacity Building, supporting the development of science-based research and regulatory programs and promoting global food security, particularly in developing and newly emergent economies;
- Biotechnology Activities With International Organizations, utilizing a variety of approaches like briefings staged at multinational meetings (e.g., at Codex and at the Rome World Food Summit in 2002) and at regional workshops.

FAS generally convenes a USDA interagency team weekly to review SPS and TBT problems in agricultural trade, including those in GE products; is involved in developing policy on matters requiring higher-level decisions; and leads or accompanies other agency officials on overseas missions to discuss and resolve issues.

FAS staff have long been engaged in biotechnology trade issues, but generally not on a full-time basis. Seeking to improve its focus and work, the agency in 2002 established a new Biotechnology Office reporting to the Administrator and Associate Administrator. The office hopes to have 10 full-time staffers on board by August 2003 organized around four teams devoted exclusively to biotechnology work: bilateral relations in Asia and the Americas; bilateral relations in Europe, the Middle East, and Africa; international organization work (e.g., with WTO, FAO, etc.); and one devoted to technical assistance, capacity building, and outreach.

USDA's **Animal & Plant Health Inspection Service (APHIS)** regulates the movement, testing, and commercialization of new GE crop varieties (and GE veterinary biologies) in the United States. In addition, as the lead federal agency regulating commodity imports and exports, APHIS works to minimize trade disruptions caused by animal and plant health issues, including those related to biotechnology. APHIS staff participate, and sometimes take the lead, in U.S. negotiations with trading partners and in multinational forums to resolve SPS-related disputes and to harmonize international standards, to ensure that such issues are not used as unfair barriers to trade.

An increasing share of APHIS harmonization work revolves around trade in GE products. The agency has sought both to ensure that international biotechnology policies are science-based, and to promote the global credibility of the U.S. GE regulatory process. It is involved in Codex, IPPC, OIE, the Cartagena Protocol, and the OECD, among other organizations, and is in ongoing negotiations with Canada on harmonization, which led recently to a draft bilateral agreement on environmental assessment criteria for transgenic plants. In FY2002, APHIS consolidated its biotechnology personnel under a single new office, Biotechnology Regulatory Services, covering both domestic regulatory activities; and international trade, policy, and capacity building. The office was assigned 46 staff slots in FY2002.[26]

The Federal Grain Inspection Service of USDA's **Grain Inspection, Packers & Stockyards Administration (GIPSA)** spells out official grain quality standards, establishes standard testing methodologies, and, for a fee, tests grain (a requirement for most exports) to ensure they meet such standards. GIPSA has articulated a biotechnology program in response to the "emergence of value-enhanced grains and oilseeds, development of niche markets for non-biotech commodities, and establishment of new regulatory requirements by U.S. trading partners [which] has created a need for greater product differentiation in the marketplace."[27]

GIPSA has been evaluating the performance of tests to detect the presence of GMOs in grains and oilseeds and offering a proficiency program for organizations testing for biotechnology-derived grains and oilseeds to improve reliability. The agency also has developed sampling guidelines for the industry, among other

[26] USDA. *2004 Budget Explanatory Notes.*

[27] GIPSA. *GIPSA's Biotechnology Program,* which can be viewed at [http://www.usda.gov/ gipsa/ newsroom/backgroundersAjiobackgrounder.htm].

services. GIPSA itself has not been testing for GMOs in shipments, except for the presence of StarLink™ corn, as a voluntary fee-based service.[28]

[28] In 2001, StarLink™, a GE corn approved only for animal and not human food, was found in taco shells. See CRS Report RS20732, StarLink™ Corn Controversy: Background.

INDEX

T

U

V

W

Y

Z